THE WORDS THAT RESHAPED AMERICA

F·D·R

THE WORDS THAT RESHAPED AMERICA

F·D·R

Stamford Parker

Quill
An Imprint of HarperCollins*Publishers*

First Quill edition published 2000

Library of Congress Control Number: 00-132687

ISBN: 0-380-80070-5

00 01 02 03 04 RRD 10 9 8 7 6 5 4 3 2 1

THE WORDS THAT RESHAPED AMERICA

F·D·R

INTRODUCTION

On March 8, 1933, Franklin Delano Roosevelt held a press conference unlike any that had preceded it. He said to the reporters gathered around his desk: "I am told that what I am about to do will become impossible, but I am going to try it. We are not going to have anymore written questions." And discarding those presubmitted by the press, he invited the reporters to ask him anything they wanted to. This candor, this willingness to tell the truth and tell it straight, was one of Roosevelt's greatest strengths as a speaker and a writer.

He also possessed total confidence in himself and his ideas. What he would say about his presidential run of 1936 would hold true for much of what he sought to do in office: "There's one issue in this campaign. It's myself, and people must be either for me or against me." With the unquenchable passion and supreme eloquence born of such confidence, he ensured that most Americans chose the former stance.

Both his candor and his confidence, though, were enabled by his absolute faith in the people he led.

He firmly believed Americans could handle even the most difficult news. They didn't need to be coddled by spin or mesmerized by high-born rhetoric. As he would say to his speechwriter, Samuel Rosenman, after Pearl Harbor, "I want to explain the war in laymen's language; if they understand the problem and what we are driving at, I am sure that they can take any kind of bad news right on the chin." (One could easily replace "the war" with "the Depression.")

And if he exhorted Americans in speech after speech to hope, even when salvation seemed impossibly distant, he also told them they could bring about that salvation themselves. For example, in his Fireside Chat on the National Recovery Administration, he said, "When Andrew Jackson, 'Old Hickory,' died, someone asked, 'Will he go to heaven?' and the answer was, 'He will if he wants to.' If I am asked whether the American people will pull themselves out of this depression, I answer, 'They will if they want to.' " If there is any kind of doubt in this, FDR banished it within seconds: "I cannot guarantee the success of this nationwide plan, but the people of this country can guarantee its success." And how could he make such a guarantee? Because "disorder is not an American habit," he said in his first annual message to Congress. "Self-help and self-control are the essence of the American tradition—not of necessity the form of that tradition, but its spirit. The program itself comes from the American people."

Most importantly, Roosevelt reminded Americans constantly that he was there in the trenches with them, backing up his fine words with even finer deeds, giving Americans every chance to succeed. We see this in his first Fireside Chat, in which he tried to prevent runs on the banks. He closed with: "You people must have faith; you must not be stampeded by rumors or guesses. Let us unite in banishing fear. We have provided the machinery to restore our financial system; it is up to you to support and make it work. It is your problem no less than it is mine. Together we cannot fail." And we see it again in his third inaugural address. As Hitler marched across Europe, Roosevelt ended by saying, "In the face of great perils never before encountered, our strong purpose is to protect and to perpetuate the integrity of democracy. For this we muster the spirit of America, and the faith of America. We do not retreat. We are not content to stand still. As Americans, we go forward, in the service of our country, by the will of God."

Roosevelt's candor and confidence, his faith in his people, his swift and decisive actions that time and again proved his words true: these were what made his subsequent "unrehearsed" press conferences—more than 1,000 over the next twelve years—so effective. They are what made his hundreds of Fireside Chats so memorable; his four inaugural addresses, his fourteen annual messages to Congress, his countless other speeches and statements, so powerful. They are, in the end, why Roosevelt's words and their wisdom had such a profound and permanent effect on America.

sistently that he was there in the trenches with them, backing up his fine words with even finer deeds, giving Americans every chance to succeed. We see this in his first fireside chat, in which he tried to prevent a run on the banks. He closed with: "You people must have faith; you must not be stampeded by rumors or guesses. Let us unite in banishing fear. We have provided the machinery to restore our financial system; it is up to you to support and make it work. It is your problem no less than it is mine. Together we cannot fail." And we see it again in his third inaugural address. As Hitler marched across Europe, Roosevelt ended by saying: "In the face of great perils never before encountered, our strong purpose is to protect and to perpetuate the integrity of democracy. For this we muster the spirit of America, and the faith of America. We do not retreat. We are not content to stand still. As Americans, we go forward, in the service of our country, by the will of God."

Roosevelt's contagious confidence, his faith in his people, his swift and decisive actions that time and again proved his words true—these were what made his speeches [illegible] over the next twelve years—so effective. They are what made the hundreds of remarks, chats, comments, talks, his four inaugural addresses, his twelve annual messages to Congress, the countless other speeches and statements, so powerful. They are, in the end, why Roosevelt's words and their wisdom had such a profound and permanent effect on America.

In Roosevelt's first gubernatorial inaugural address, on January 1, 1929, what is remarkably clear is the degree to which FDR's personal style and substantial positivist demeanor had already evolved. His themes of a participatory population going out and getting the job done, and his view of the American public in general, were already fully established. He would return to many of the themes of this speech over and over throughout his political career.

IT IS A proud thing to be a citizen of the state of New York, not because of our great population and our natural resources, nor on account of our industries, our trade, or agricultural development, but because the citizens of this state, more than any other state in the Union, have grown to realize the interdependence on each other which modern civilization has created.

Under the leadership of the great governor whose place you have selected me to fill has come a willingness on our part to give as well as to receive, to aid, through the agency of the state, the well-being of the men and women who, by their toil, have made our material prosperity possible.

I object to having this spirit of personal civil responsibility to the state and to the individual, which has placed New York in the lead as a progressive commonwealth, described as "humanitarian." It is far more than that. It is the recognition that our civilization cannot endure unless we, as individuals, realize our personal responsibility to and dependence on the rest of the world. For it is literally true that the "self-supporting" man or woman has become as extinct as the man of the Stone Age. Without the help of thousands of others, any one of us would die, naked and starved. Consider the bread upon our table, the clothes upon our backs, the luxuries that make life pleasant; how many men worked in sunlit fields, in dark mines, in the fierce heat of molten metal, and among the looms and wheels of countless factories, in order to create them for our use and enjoyment.

I am proud that we of this state have grown to realize this dependence, and, what is more important, have also come to know that we, as individuals, in our turn must give our time and our intelligence to help those who have helped us. To secure more of life's pleasures for the farmer; to guard the toilers in the factories and ensure them a fair wage and protection from the dangers of their trades; to compensate them by adequate insurance for injuries received while working for us; to open the doors of knowledge to their children more widely; to aid those who are crippled and ill; to pursue with strict justice all evil persons who prey upon their fellow men; and at the same time, by intelligent and helpful sympathy, to lead wrongdoers into right paths—all of these great aims of life are more fully realized here than in any other state in the Union. We have but started on the road, and we have far to go; but during the last six years in particular, the people of this state have shown their impatience of those who seek to make such things a football of politics or by blind, unintelligent obstruction attempt to bar the road to progress.

Most gratifying of all, perhaps, is the practical way in which we have set about to take the first step toward this higher civilization, for, first of all, has been the need to set our machinery of government in order. If we are to reach these aims efficiently without needless waste of time or money, we must continue the efforts to simplify and modernize. You cannot build a modern dynamo with the ancient forge and bellows of the medieval blacksmith. The modernization of our administrative procedure, not alone that of the state, but also of those other vital units of countries, of cities, of towns, and of villages, must be accomplished; and while in the unit of the state we have almost reached our goal, I want to emphasize that in other units we have a long road to travel.

Each one of us must realize the necessity of our personal interest, not only toward our fellow citizens, but in the government itself. You must watch, as a public duty, what is done and what is not done at Albany. You must understand the issues that arise in the legislature, and the recommendations made by your governor, and judge for yourselves if they are right or wrong. If you find them right it is your duty as citizens on next election day to repudiate those who oppose, and to support by your vote those who strive for their accomplishment.

I want to call particularly on the public press of this state, in whose high standards I have the greatest confidence, to devote more space to the explanation and consideration of such legislation as may come up this year, for no matter how willing the individual citizen may be to support wise and progressive measures, it is only through the press—and I mean not only our great dailies but their smaller sisters in the rural district—that our electorate can learn and understand what is going on.

There are many puzzling problems to be solved. I shall here mention but three. In the brief time that I have been speaking to you, there has run to waste on their paths toward the sea enough power from our rivers to have turned the wheels of a thousand factories, to have lit a million farmers' homes—power which nature has supplied us through the gift of God. It is intolerable that the utilization of this stupendous heritage should be longer delayed by petty squabbles and partisan dispute. Time will not solve the problem; it will be more difficult as time goes on to reach a fair conclusion. It must be solved now.

I should like to state clearly the outstanding features of the problem itself. First, it is agreed, I think, that the water power of the state should belong to all the people. There was, perhaps, some excuse for careless legislative gift of power sites in the days when it was of no seemingly great importance. There can be no such excuse now. The title to this power must vest forever in the people of this state. No commission—no, not the legislature itself has any right to give, for any consideration whatever, a single potential kilowatt in virtual perpetuity to any person or corporation whatsoever. The legislature in this matter is but the trustee of the people, and it is its solemn duty to administer such heritage so as most greatly to benefit the whole people. On this point there can be no dispute.

It is also the duty of our legislative bodies to see that this power, which belongs to all the people, is transformed into usable electrical energy and distributed to them at the lowest possible cost. It is our power; and no inordinate profits must be allowed to those who act as the people's agents in bringing this power to their homes and workshops. If we keep these two fundamental facts before us, half of the problem disappears.

There remains the technical question as to which of several methods will bring this power to our doors with the least expense. Let me here make clear the three divisions of this technical side of the question.

First, the construction of the dams, the erection of power houses and the installation of the turbines necessary to convert the force of the falling water into electricity;

Second, the construction of many thousands of miles of transmission lines to bring the current so produced to the smaller distributing centers throughout the state; and

Third, the final distribution of this power into thousands of homes and factories.

How much of this shall be undertaken by the state, how much of this carried out by properly regulated private enterprises, how much of this by some combination of the two, is the practical question that we have before us. And in the consideration of the question I want to warn the people of this state against too hasty assumption that mere regulation by public service commissions is, in itself, a sure guarantee of protection of the interest of the consumer.

The questionable taking of jurisdiction by federal courts, the gradual erection of a body of court-made law, the astuteness of our legal brethren, the possible temporary capitulation of our public servants and even of a dormant public opinion itself may, in the future, as in the past, nullify the rights of the public.

I, as your governor, will insist—and I trust with the support of the whole people—that there be no alienation of our possession of and title to our power sites, and that whatever method of distribution be adopted, there be no possible legal thwarting of the protection of the people themselves from excessive profits on the part of anybody.

On another matter I tread perhaps a new path. The phrase "rich man's justice" has become too common nowadays. So complicated has our whole legal machinery become through our attempt to mend antiquated substructures by constant patching of the legal procedure and the courts that justice is our most expensive commodity. That rich criminals too often escape punishment is a general belief of our people. The difficulty with which our citizens maintain their civil rights before the courts has not been made a matter of such public notice but is equally serious. It is my hope that within the next two years we shall have begun to simplify and to cheapen justice for the people.

Lastly, I want to refer to the difficult situation to which in recent years a large part of the rural population of our state has come. With few exceptions it has not shared in the prosperity of the urban centers.

It is not enough to dismiss this problem with the generality that it is the result of changing economic conditions. It is time to take practical steps to relieve our farm population of unequal tax burdens, to install economies in the methods of local government, to devise sounder marketing to stabilize what has been too much a speculative industry, and finally to encourage the use of each acre of our state for the purpose to which it is by nature most suited. I am certain that the cities will cooperate to this end, and that, more and more, we as citizens shall become state-minded.

May I, as your newly elected governor, appeal for your help, for your advice, and, when you feel it is needed, for your criticism? No man may be a successful governor without the full assistance of the people of his own commonwealth.

Were I as wise as Solomon, all that I might propose or decide would be mere wasted effort, unless I have your constant support. On many of the great state questions that confront us, the platforms and the public pledges of candidates of both parties are substantially agreed. We have passed through a struggle against old-time political ideas, against antiquated conservatism, against ignorance of modern conditions, marked by serious disagreements between the legislative and the executive branches of the government. As I read the declarations of both parties in asking the support of the people at the polls, I can see little reason for further controversies of this kind.

There is a period in our history known in all our school books as the Era of Good Feeling. It is my hope that we stand on the threshold of another such era in this state. For my part, I pledge that the business of the state will not be allowed to become involved in partisan politics and that I shall not attempt to claim unfair advantage for my party or for myself, for the accomplishing of those things on which we are all agreed.

You have honored me greatly by selecting me as your chief executive. It is my hope that I shall not fail you in this critical period of our history. I wish that you may have a continuance of good government and the happiest of New Years.

He was soon "honored greatly" again. Here is Roosevelt's second gubernatorial inaugural address, delivered on January 1, 1931.

TO ALL OF you who are here today and to all the people of the state of New York I extend New Year's Greetings. May 1931 be a year of happiness, a year of greater well-being, a year in which all of us may dedicate ourselves more unselfishly and more truly to the good of our commonwealth and of our fellow men.

Twenty years ago today I first entered into public service, and on this anniversary it is not unnatural that I should think of the progress of government during that period. On January 1, 1911, the people of New York were experiencing for the first time in many decades a sustained public interest in their state government, an interest first stimulated by the fine insistence of Governor Hughes and later translated into action at several legislative sessions. At that time were laid the foundations for the continuing general attention of our citizens toward state government which has followed through all these years.

With this awakening of interest followed logically the studies of the structure and functioning of the state government itself, which culminated in the Constitutional Convention of 1915 and eventually resulted in the reorganization of the departments and the creation of the present businesslike budget system under the leadership of Governor Smith.

As a summing up of this quarter century I think that we can well say that we have made great strides in modernizing the government of the state and in vastly increasing both its honesty and its efficiency; in bringing this government into a sound and responsible relationship to the social needs and the welfare of the citizens themselves; and at the same time in avoiding the pitfalls of paternalism.

This noteworthy and continuing improvement in the theory and practice of state government has come in part from progressive leadership, but most of all from a genuine public interest backed up by the willingness of thousands of citizens to give practical and unselfish service. Therefore while properly we recognize the many tasks, the many new problems which lie ahead, still I think we can take genuine pride and satisfaction in the structure and functioning of that instrumentality of our sovereignty known as the government of the state of New York.

But this gratifying modernization and perfecting of our state government serve at the same time to accentuate by contrast our lack of progress in improving local government. Not long ago I received a letter from an eminent editor, telling me tearfully that all local government had broken down, and begging me as governor—note the unconscious willingness to accept a Czar or absolute dictator in the governor's chair in Albany—to usurp and assume the functions of the officials duly elected by the communities themselves. He ended with the suggestion that if I did not do so, the alternative would be to call out the militia and establish martial law.

I cite this as an illustration of the present dangerous tendency to forget a fundamental of American democracy, which rests on the right of a locality to manage its local affairs—the tendency to encourage concentration of power at the top of a governmental structure alien to our system and more closely akin to a dictatorship or the central committee of a communistic regime.

Now my friend was right about one thing. Local government as a whole is open to severe criticism, and this applies to the cities, to the towns, to the counties, and to the multiplicity of other local agencies. During this century the problem of local government has been complicated by three new factors: the unparalleled growth of city populations, the birth of a new type of community known as the suburban area, and, piled on top of these, a wholly new series of human physical needs such as highways, pavements, water, sewers, lighting, bridges, tunnels, and schools, which affect every inhabited area of the state. We have shifted and shuffled our population by the millions, we have made more physical changes in 25 years than in the previous 250, and we have expected the old machinery of local government without any redesigning to carry the new load. The only real wonder is that the inefficiency is not worse, and that public honesty is probably on a higher general plane among the rank and file than it was a generation ago.

But why are our local governments archaic in design, unsuited for the purpose for which they are established, unsatisfactory in their functioning, and profligate in the spending of the taxpayers' contributions? The answer is not hard to find. It is because the individual citizen is indifferent to his local government problems. The stress of business competition in this hectic twentieth century of ours, the even more feverish pursuit of pleasure to compensate for our strenuous business days—these so occupy the time and thought of our average taxpayer as to leave no inclination either to study or assist in the conduct of the community in which he lives. We do not trust our personal business affairs to strangers; we do not take our pleasures vicariously; but when it comes to running our local communities we gladly let John Doe do it. We do not even take the trouble to inquire what manner of person John Doe may be when he is nominated by a political party for a responsible local office. We do not know enough about the machinery of our local government to find out after he is elected whether he has been an efficient official or not. We know neither what his job is nor how he has performed it. We grumble at the taxes, but have not the slightest idea what could be done to lower them, because we do not know for what they are spent. We are occasionally aroused into driving out the grafter and the crook but we allow complacently a hundred times the amount of their peculations to be frittered away for needless and costly duplication of governmental functions under a system designed originally for the simple needs of our colonial forefathers. We are far more familiar with the structure of our national government at Washington and of our state government at Albany than we are with the government of our own town, village, city, or county. Our criticism is seldom valuable because we have no clear idea of what is wrong and not the faintest real conception of any practical remedy.

It would be appalling to compute the millions of dollars which our local taxpayers needlessly pay as the price of their indifference. Our state government has been reformed because of the ability of great leaders to awaken public interest in the affairs of state. Our local communities, alas, only in rare instances, have aroused themselves to an interest in local government extending down to the average citizen.

Too often reform associations, preachers, editors, committees of eminent citizens have, with the best intentions, come forward for a few months only to right an emergency evil, or else they have represented one angle only in a many-sided problem, or else they have had an ulterior motive of mere partisanship. It is impossible to get better government locally if only 5 percent of the electorate participate in civic affairs two months of the year while the other 95 percent remain uninterested all twelve months of the year.

Let us get this picture clearly before us. We have three main divisions of government: first, the federal, operating, at least in theory, in the national field and in accordance with strictly limited powers of control ceded to it by the sovereign states; second, the state, moving only in the statewide field and careful not to confuse state with local functions; third, the several divisions of local government, authorized to be sure by the state, but based on the centuries-old conception of the town meeting.

When states become indifferent to their duties, the natural tendency is for the national government to grasp for more power. In like manner if community government becomes slipshod and lax, the way of opportunism is to carry the problem to Albany and get the local legislators to introduce a bill.

Some of us for many years have striven against the multiplicity of local laws passed each year by complacent legislatures. During the past two years I have the record of having vetoed more bills of this character than any previous governor. Back of the immediate reason in the individual case has been the fundamental thought that the elimination of piecemeal legislation might to some degree make people think along broader lines and come back another year with a well-thought-out comprehensive plan to cover the subject as a whole. This is, of course, in direct line with what is commonly called the home rule principle. The principle is right. We have started to apply it. But at the same time its success must depend on local interest on the part of the 95 percent of the citizens as well as of the 5 percent if we are not to slip back into the system of a generation ago when even the number of dog catchers in a community was made the subject of state legislation.

How often do people stop to realize the relative influence which the three great branches of government play in their daily lives? What daily contact has the average man or woman with the activities, the machinery, or the administration of the government in Washington? What personal touch has the average man or woman with the government in Albany except when traveling over a state highway or visiting a state park? If you own a piece of real estate, whether it be a business property or a home or a farm, do you ever stop to realize that when you pay the taxes on it not one cent goes to Albany or to Washington?

Much has been said of the mounting burden of taxation; yet lack of thought leads most of us to forget that by far the greater part of the increase has been caused during this generation by mounting local expenditures. And this is further emphasized by the fact that over half of all the tax money collected by the state of New York is spent not by the state, but is returned by the state to the counties and communities for local expenditure.

It is my ardent hope that in the twenty years to come the people of New York will be able to accomplish as much for the cause of improved local government along American lines as we have accomplished for our state government in the past twenty years. This can come only through leadership and through a greater dedication on the part of the individual, giving more generously of his time and thought and personal service.

First, we need to plan, to go to the root of the problem in order that specific remedies may be offered. When this is done I am very certain that the legislature and the administration of this state will lend their aid and sympathy to the granting of the necessary authority.

I have spent two happy years in Albany and today I look forward to 1931 and 1932 as giving me, because of the experience which I have had, a greater opportunity for usefulness to the people of the state. The older I grow the more insistent I am that the average of the citizenship of the state will respond to the presentation of problems of government if the fundamentals of the issues are presented to them clearly and honorably. I am convinced, too, that the electorate has a sense of proportion and, given the facts, will of its own accord recognize them in the order of their importance.

That is why on this New Year's Day I have taken occasion to speak somewhat at length on government. One part at least of our government we can be proud of—the state is functioning well. Another part, the one which enters more closely than any other into our daily affairs, needs our attention, our interest, and our earnest efforts to improve.

The vitality of our American institutions has been amply proven in the past. We have met difficulties before this and have solved them in accordance with the basic theories of a representative democracy. Let us not at this time pursue the easy road of centralization of authority lest someday we discover too late that our liberties have disappeared. Let us pause in our pursuit of materialism and pleasure and devote greater efforts to retain these liberties within the communities in which we dwell.

That is the New Year's message from the governor of your state who pledges himself anew to devote the best that is in him to the wholehearted service of the men and women and children of the state of New York.

His so-called "Forgotten Man" address, a radio address on the economic emergency, was delivered April 7, 1932; presaging the concerns he would address more fully at the Democratic Convention later that year.

FIFTEEN YEARS AGO, my public duty called me to an active part in a great national emergency, the World War. Success then was due to a leadership whose vision carried beyond the timorous and futile gesture of sending a tiny army of 150,000 trained soldiers and the regular navy to the aid of our allies. The generalship of that moment conceived of a whole nation mobilized for war: Economic, industrial, social, and military resources gathered into a vast unit capable of and actually in the process of throwing into the scales 10 million men equipped with physical needs and sustained by the realization that behind them were the united efforts of 110 million human beings. It was a great plan because it was built from bottom to top and not from top to bottom.

In my calm judgment, the nation faces today a more grave emergency than in 1917.

It is said that Napoleon lost the battle of Waterloo because he forgot his infantry—he staked too much upon the more spectacular but less substantial cavalry. The present administration in Washington provides a close parallel. It has either forgotten or it does not want to remember the infantry of our economic army.

These unhappy times call for the building of plans that rest upon the forgotten, the unorganized, but the indispensable units of economic power, the plans like those of 1917 that build from the bottom up and not from the top down, that put their faith once more in the forgotten man at the bottom of the economic pyramid.

Obviously, these few minutes tonight permit no opportunity to lay down the ten or a dozen closely related objectives of a plan to meet our present emergency, but I can draw a few essentials, a beginning in fact, of a planned program.

It is the habit of the unthinking to turn in times like this to the illusions of economic magic. People suggest that a huge expenditure of public funds by the federal government and by state and local governments will completely solve the unemployment problem. But it is clear that even if we could raise many billions of dollars and find definitely useful public works to spend these billions on, even all that money would not give employment to the seven million or ten million people who are out of work. Let us admit frankly that it would be only a stopgap. A real economic cure must go to the killing of the bacteria in the system rather than to the treatment of external symptoms.

How much do the shallow thinkers realize, for example, that approximately one-half of our whole population, fifty or sixty million people, earn their living by farming or in small towns whose existence immediately depends on farms? They have today lost their purchasing power. Why? They are receiving for farm products less than the cost to them of growing these farm products. The result of this loss of purchasing power is that many other millions of people engaged in industry in the cities cannot sell industrial products to the farming half of the nation. This brings home to every city worker that his own employment is directly tied up with the farmer's dollar. No nation can long endure half bankrupt. Main Street, Broadway, the mills, the mines will close if half the buyers are broke.

I cannot escape the conclusion that one of the essential parts of a national program of restoration must be to restore purchasing power to the farming half of the country. Without this the wheels of railroads and of factories will not turn.

Closely associated with this first objective is the problem of keeping the home-owner and the farm-owner where he is, without being dispossessed through the foreclosure of his mortgage. His relationship to the great banks of Chicago and New York is pretty remote. The $2 billion fund which President Hoover and the Congress have put at the disposal of the big banks, the railroads, and the corporations of the nation is not for him.

His is a relationship to his little local bank or local loan company. It is a sad fact that even though the local lender in many cases does not want to evict the farmer or home-owner by foreclosure proceedings, he is forced to do so in order to keep his bank or company solvent. Here should be an objective of government itself, to provide at least as much assistance to the little fellow as it is now giving to the large banks and corporations. That is another example of building from the bottom up.

One other objective closely related to the problem of selling American products is to provide a tariff policy based upon economic common sense rather than upon politics, hot air, and pull. This country during the past few years, culminating with the Hawley-Smoot Tariff in 1929, has compelled the world to build tariff fences so high that world trade is decreasing to the vanishing point. The value of goods internationally exchanged is today less than half of what it was three or four years ago.

Every man and woman who gives any thought to the subject knows that if our factories run even eighty percent of capacity, they will turn out more products than we as a nation can possibly use ourselves. The answer is that if they run on eighty percent of capacity, we must sell some goods abroad. How can we do that if the outside nations cannot pay us in cash? And we know by sad experience that they cannot do that. The only way they can pay us is in their own goods or raw materials, but this foolish tariff of ours makes that impossible.

What we must do is this: revise our tariff on the basis of a reciprocal exchange of goods, allowing other nations to buy and to pay for our goods by sending us such of their goods as will not seriously throw any of our industries out of balance, and incidentally making impossible in this country the continuance of pure monopolies which cause us to pay excessive prices for many of the necessities of life.

Such objectives as these three—restoring farmers' buying power, relief to the small banks and home-owners, and a reconstructed tariff policy—are only a part of ten or a dozen vital factors. But they seem to be beyond the concern of a national administration which can think in terms only of the top of the social and economic structure. It has sought temporary relief from the top down rather than permanent relief from the bottom up. It has totally failed to plan ahead in a comprehensive way. It has waited until something has cracked and then at the last moment has sought to prevent total collapse.

It is high time to get back to fundamentals. It is high time to admit with courage that we are in the midst of an emergency at least equal to that of war. Let us mobilize to meet it.

FDR would give a huge boost to the mobilization effort, on July 2, 1932, at the Democratic Convention in Chicago. In one of the most brilliant political and economic maneuvers in Ameri-

can history, he fully fleshed out The New Deal in a long and fascinating speech.

❧ CHAIRMAN WALSH, MY friends of the Democratic National Convention of 1932: I appreciate your willingness after these six arduous days to remain here, for I know well the sleepless hours which you and I have had. I regret that I am late, but I have no control over the winds of heaven and could only be thankful for my navy training.

The appearance before a national convention of its nominee for president, to be formally notified of his selection, is unprecedented and unusual, but these are unprecedented and unusual times. I have started out on the tasks that lie ahead by breaking the absurd traditions that the candidate should remain in professed ignorance of what has happened for weeks until he is formally notified of that event many weeks later.

My friends, may this be the symbol of my intention to be honest and to avoid all hypocrisy or sham, to avoid all silly shutting of the eyes to the truth in this campaign. You have nominated me and I know it, and I am here to thank you for the honor.

Let it also be symbolic that in so doing I broke traditions. Let it be from now on the task of our party to break foolish traditions. We will break foolish traditions and leave it to the Republican leadership, far more skilled in that art, to break promises.

Let us now and here highly resolve to resume the country's interrupted march along the path of real progress, of real justice, of real equality for all of our citizens, great and small. Our indomitable leader in that interrupted march is no longer with us, but there still survives today his spirit. Many of his captains, thank God, are still with us, to give us wise counsel. Let us feel that in everything we do there still lives with us, if not the body, the great indomitable, unquenchable, progressive soul of our commander in chief, Woodrow Wilson.

I have many things on which I want to make my position clear at the earliest possible moment in this campaign. That admirable document, the platform which you have adopted, is clear. I accept it 100 percent.

And you can accept my pledge that I will leave no doubt or ambiguity on where I stand on any question of moment in this campaign.

As we enter this new battle, let us keep always present with us some of the ideals of the party: the fact that the Democratic party by tradition and by the continuing logic of history, past and present, is the bearer of liberalism and of progress and at the same time of safety to our institutions. And if this appeal fails, remember well, my friends, that a resentment against the failure of Republican leadership—and note well that in this campaign I shall not use the words "Republican party," but I shall use, day in and day out, the words "Republican leadership"—the failure of Republican leaders to solve our troubles may degenerate into unreasoning radicalism.

The great social phenomenon of this depression, unlike others before it, is that it has produced but a few of the disorderly manifestations that too often attend upon such times.

Wild radicalism has made few converts, and the greatest tribute that I can pay to my countrymen is that in these days of crushing want there persists an orderly and hopeful spirit on the part of the millions of our people who have suffered so much. To fail to offer them a new chance is not only to betray their hopes but to misunderstand their patience.

To meet by reaction that danger of radicalism is to invite disaster. Reaction is no barrier to the radical. It is a challenge, a provocation. The way to meet that danger is to offer a workable program of reconstruction, and the party to offer it is the party with clean hands.

This, and this only, is a proper protection against blind reaction on the one hand and an improvised, hit-or-miss, irresponsible opportunism on the other.

There are two ways of viewing the government's duty in matters affecting economic and social life. The first sees to it that a favored few are helped and hopes that some of their prosperity will leak through, sift through, to labor, to the farmer, to the small-businessman. That theory belongs to the party of Toryism, and I had hoped that most of the Tories left this country in 1776.

But it is not and never will be the theory of the Democratic party. This is no time for fear, for reaction or for timidity. Here and now I invite those nominal Republicans who find that their conscience cannot be squared with the groping and the failure of their party leaders to join hands with us; here and now, in equal measure, I warn those nominal Democrats who squint at the future with their faces turned toward the past, and who feel no responsibility to the demands of the new time, that they are out of step with their party.

Yes, the people of this country want a genuine choice this year, not a choice between two names for the same reactionary doctrine. Ours must be a party of liberal thought, of planned action, of enlightened international outlook, and of the greatest good to the greatest number of our citizens.

Now it is inevitable—and the choice is that of the times—it is inevitable that the main issue of this campaign should revolve about the clear fact of our economic condition, a depression so deep that it is without precedent in modern history. It will not do merely to state, as do Republican leaders to explain their broken promises of continued inaction, that the depression is worldwide. That was not their explanation of the apparent prosperity of 1928. The people will not forget the claim made by them then that prosperity was only a domestic product manufactured by a Republican president and a Republican Congress. If they claim paternity for the one they cannot deny paternity for the other.

I cannot take up all the problems today. I want to touch on a few that are vital. Let us look a little at the recent history and the simple economics, the kind of economics that you and I and the average man and woman talk.

In the years before 1929 we know that this country had completed a vast cycle of building and inflation; for ten years we expanded on the theory of repairing the wastes of the war, but actually expanding far beyond that, and also beyond our natural and normal growth. Now it is worth remembering, and the cold figures of finance prove it, that during that time there was little or no drop in the prices that the consumer had to pay, although those same figures proved that the cost of production fell very greatly; corporate profit resulting from this period was enormous at the same time little of that profit was devoted to the reduction of prices. The consumer was forgotten. Very little of it went into increased wages; the worker was forgotten, and by no means an adequate proportion was even paid out in dividends—the stockholder was forgotten.

And, incidentally, very little of it was taken by taxation to the beneficent government of those years.

What was the result? Enormous corporate surpluses piled up—the most stupendous in history. Where, under the spell of delirious speculation, did those surpluses go? Let us talk economics that the figures prove and that we can understand. Why, they went chiefly in two directions: first, into new and unnecessary plants which now stand stark and idle; and second, into the call-money market of Wall Street, either directly by the corporations, or indirectly through the banks. Those are the facts. Why blink at them?

Then came the crash. You know the story. Surpluses invested in unnecessary plants became idle. Men lost their jobs; purchasing power dried up; banks became frightened and started calling loans. Those who had money were afraid to part with it. Credit contracted. Industry stopped. Commerce declined, and unemployment mounted.

And there we are today.

Translate that into human terms. See how the events of the past three years have come home to specific groups of people: first, the group dependent on industry; second, the group dependent on agriculture; third, and made up in large part of members of the first two groups, the people who are called "small investors and depositors." In fact, the strongest possible tie between the first two groups, agriculture and industry, is the fact that the savings and to a degree the security of both are tied together in that third group—the credit structure of the nation.

Never in history have the interests of all the people

been so united in a single economic problem. Picture to yourself, for instance, the great groups of property owned by millions of our citizens, represented by credits issued in the form of bonds and mortgages—government bonds of all kinds, federal, state, county, municipal; bonds of industrial companies, of utility companies; mortgages on real estate in farms and cities, and finally the vast investments of the nation in the railroads. What is the measure of the security of each of those groups? We know well that in our complicated, interrelated credit structure if any one of these credit groups collapses they may all collapse. Danger to one is danger to all.

How, I ask, has the present administration in Washington treated the interrelationship of these credit groups? The answer is clear: it has not recognized that interrelationship existed at all. Why, the nation asks, has Washington failed to understand that all of these groups, each and every one, the top of the pyramid and the bottom of the pyramid, must be considered together, that each and every one of them is dependent on every other; each and every one of them affecting the whole financial fabric?

Statesmanship and vision, my friends, require relief to all at the same time.

Just one word or two on taxes, the taxes that all of us pay toward the cost of government of all kinds.

I know something of taxes. For three long years I have been going up and down this country preaching that government—federal and state and local—costs too much. I shall not stop that preaching. As an immediate program of action we must abolish useless offices. We must eliminate unnecessary functions of government—functions, in fact, that are not definitely essential to the continuance of government. We must merge, we must consolidate subdivisions of government, and, like the private citizen, give up luxuries which we can no longer afford.

By our example at Washington itself, we shall have the opportunity of pointing the way of economy to local government, for let us remember well that out of every tax dollar in the average state in this nation, forty cents enter the Treasury in Washington, D.C., ten or twelve cents only go to the state capitals, and forty-eight cents are consumed by the costs of local government in counties and cities and towns.

I propose to you, my friends, and through you, that government of all kinds, big and little, be made solvent and that the example be set by the president of the United States and his Cabinet.

And talking about setting a definite example, I congratulate this convention for having had the courage fearlessly to write into its declaration of principles what an overwhelming majority here assembled really thinks about the Eighteenth Amendment. This convention wants repeal. Your candidate wants repeal. And I am confident that the United States of America wants repeal.

Two years ago the platform on which I ran for governor the second time contained substantially the same provision. The overwhelming sentiment of the people of my state, as shown by the vote of that year, extends, I know, to the people of many of the other states. I say to you now that from this date on the Eighteenth Amendment is doomed. When that happens, we as Democrats must and will, rightly and morally, enable the states to protect themselves against the importation of intoxicating liquor where such importation may violate their state laws. We must rightly and morally prevent the return of the saloon.

To go back to this dry subject of finance, because it

all ties in together—the Eighteenth Amendment has something to do with finance, too—in a comprehensive planning for the reconstruction of the great credit groups, including government credit, I list an important place for that prize statement of principle in the platform here adopted calling for the letting in of the light of day on issues of securities, foreign and domestic, which are offered for sale to the investing public.

My friends, you and I as common-sense citizens know that it would help to protect the savings of the country from the dishonesty of crooks and from the lack of honor of some men in high financial places. Publicity is the enemy of crookedness.

And now one word about unemployment, and incidentally about agriculture. I have favored the use of certain types of public works as a further emergency means of stimulating employment and the issuance of bonds to pay for such public works, but I have pointed out that no economic end is served if we merely build without building for a necessary purpose. Such works, of course, should insofar as possible be self-sustaining if they are to be financed by the issuing of bonds. So as to spread the points of all kinds as widely as possible, we must take definite steps to shorten the working day and the working week.

Let us use common sense and business sense. Just as one example, we know that a very hopeful and immediate means of relief, both for the unemployed and for agriculture, will come from a wide plan of the converting of many millions of acres of marginal and unused land into timberland through reforestation. There are tens of millions of acres east of the Mississippi River alone in abandoned farms, in cut-over land, now growing up in worthless brush. Why, every European nation has a definite land policy, and has had one for generations. We have none. Having none, we face a future of soil erosion and timber famine. It is clear that economic foresight and immediate employment march hand in hand in the call for the reforestation of these vast areas.

In so doing, employment can be given to a million men. That is the kind of public work that is self-sustaining, and therefore capable of being financed by the issuance of bonds which are made secure by the fact that the growth of tremendous crops will provide adequate security for the investment.

Yes, I have a very definite program for providing employment by that means. I have done it, and I am doing it today in the state of New York. I know that the Democratic party can do it successfully in the nation. That will put men to work, and that is an example of the action that we are going to have.

Now as a further aid to agriculture, we know perfectly well—but have we come out and said so clearly and distinctly?—we should repeal immediately those provisions of law that compel the federal government to go into the market to purchase, to sell, to speculate in farm products in a futile attempt to reduce farm surpluses. And they are the people who are talking of keeping government out of business. The practical way to help the farmer is by an arrangement that will, in addition to lightening some of the impoverishing burdens from his back, do something toward the reduction of the surpluses of staple commodities that hang on the market. It should be our aim to add to the world prices of staple products the amount of a reasonable tariff protection, to give agriculture the same protection that industry has today.

And in exchange for this immediately increased return

I am sure that the farmers of this nation would agree ultimately to such planning of their production as would reduce the surpluses and make it unnecessary in later years to depend on dumping those surpluses abroad in order to support domestic prices. That result has been accomplished in other nations; why not in America, too?

Farm leaders and farm economists, generally, agree that a plan based on that principle is a desirable first step in the reconstruction of agriculture. It does not in itself furnish a complete program, but it will serve in great measure in the long run to remove the pall of a surplus without the continued perpetual threat of world dumping. Final voluntary reduction of surplus is a part of our objective, but the long continuance and the present burden of existing surpluses make it necessary to repair great damage of the present by immediate emergency measures.

Such a plan as that, my friends, does not cost the government any money, nor does it keep the government in business or in speculation.

As to the actual wording of a bill, I believe that the Democratic party stands ready to be guided by whatever the responsible farm groups themselves agree on. That is a principle that is sound; and again I ask for action.

One more word about the farmer, and I know that every delegate in this hall who lives in the city knows why I lay emphasis on the farmer. It is because one-half of our population, over fifty million people, are dependent on agriculture; and, my friends, if those fifty million people have no money, no cash, to buy what is produced in the city, the city suffers to an equal or greater extent.

That is why we are going to make the voters understand this year that this nation is not merely a nation of independence, but it is, if we are to survive, bound to be a nation of interdependence—town and city, and North and South, East and West. That is our goal, and that goal will be understood by the people of this country no matter where they live.

Yes, the purchasing power of that half of our population dependent on agriculture is gone. Farm mortgages reach nearly ten billions of dollars today and interest charges on that alone are $560 million a year. But that is not all. The tax burden caused by extravagant and inefficient local government is an additional factor. Our most immediate concern should be to reduce the interest burden on these mortgages.

Rediscounting of farm mortgages under salutary restrictions must be expanded and should, in the future, be conditioned on the reduction of interest rates. Amortization payments, maturities should likewise in this crisis be extended before rediscount is permitted where the mortgagor is sorely pressed. That, my friends, is another example of practical, immediate relief: action.

I aim to do the same thing, and it can be done, for the small homeowner in our cities and villages. We can lighten his burden and develop his purchasing power. Take away, my friends, that specter of too high an interest rate. Take away that specter of the due date just a short time away. Save homes; save homes for thousands of self-respecting families, and drive out that specter of insecurity from our midst.

Out of all the tons of printed paper, out of all the hours of oratory, the recriminations, the defenses, the happy-thought plans in Washington and in every state, there emerges one great, simple, crystal-pure fact that during the past ten years a nation of 120 million people has been led by the Republican leaders to erect an impregnable barbed wire entanglement around its borders through the instrumentality of tariffs which have isolated us from all the other human beings in all the rest of the round world. I accept that admirable tariff statement in the platform of this convention. It would protect American business and American labor. By our acts of the past we have invited and received the retaliation of other nations. I propose an invitation to them to forget the past, to sit at the table with us, as friends, and to plan with us for the restoration of the trade of the world.

Go into the home of the businessman. He knows what the tariff has done for him. Go into the home of the factory worker. He knows why goods do not move. Go into the home of the farmer. He knows how the tariff has helped to ruin him.

At last our eyes are open. At last the American people are ready to acknowledge that Republican leadership was wrong and that the democracy is right.

My program, of which I can only touch on these points, is based upon this simple moral principle: the welfare and the soundness of a nation depend first upon what the great mass of the people wish and need; and second, whether or not they are getting it.

What do the people of America want more than anything else? To my mind, they want two things: work, with all the moral and spiritual values that go with it; and with work, a reasonable measure of security—security for themselves and for their wives and children. Work and security—these are more than words. They are more than facts. They are the spiritual values, the true goal toward which our efforts of reconstruction should lead. These are the values that this program is intended to gain; these are the values we have failed to achieve by the leadership we now have.

Our Republican leaders tell us economic laws—sacred, inviolable, unchangeable—cause panics which no one could prevent. But while they prate of economic laws, men and women are starving. We must lay hold of the fact that economic laws are not made by nature. They are made by human beings.

Yes, when—not if—when we get the chance, the federal government will assume bold leadership in distress relief. For years Washington has alternated between putting its head in the sand and saying there is no large number of destitute people in our midst who need food and clothing, and then saying the states should take care of them, if there are. Instead of planning two and a half years ago to do what they are now trying to do, they kept putting it off from day to day, week to week, and month to month, until the conscience of America demanded action.

I say that while primary responsibility for relief rests with localities now, as ever, yet the federal government has always had and still has a continuing responsibility for the broader public welfare. It will soon fulfill that responsibility.

And now, just a few words about our plans for the next four months. By coming here instead of waiting for a formal notification, I have made it clear that I believe we should eliminate expensive ceremonies and that we should set in motion at once, tonight, my friends, the necessary machinery for an adequate presentation of the issues to the electorate of the nation.

I myself have important duties as governor of a great state, duties which in these times are more arduous and more grave than at any previous period. Yet I feel confident that I shall be able to make a number of short visits to several parts of the nation. My trips will have as their first objective the study at first hand, from the lips of men and women of all parties and all occupations, of the actual conditions and needs of every part of an interdependent country.

One word more: out of every crisis, every tribulation, every disaster, mankind rises with some share of greater knowledge, of higher decency, of purer purpose. Today we shall have come through a period of loose thinking, descending morals, an era of selfishness, among individual men and women and among nations. Blame not governments alone for this. Blame ourselves in equal share. Let us be frank in acknowledgment of the truth that many amongst us have made obeisance to Mammon, that the profits of speculation, the easy road without toil, have lured us from the old barricades. To return to higher standards we must abandon the false prophets and seek new leaders of our own choosing.

Never before in modern history have the essential differences between the two major American parties stood out in such striking contrast as they do today. Republican leaders not only have failed in material things, they have failed in national vision, because in disaster they have held out no hope, they have pointed out no path for the people below to climb back to places of security and of safety in our American life.

Throughout the nation, men and women, forgotten in the political philosophy of the government of the last years, look to us here for guidance and for more equitable opportunity to share in the distribution of national wealth.

On the farms, in the large metropolitan areas, in the

smaller cities and in the villages, millions of our citizens cherish the hope that their old standards of living and of thought have not gone forever. Those millions cannot and shall not hope in vain.

I pledge you, I pledge myself, to a New Deal for the American people. Let us all here assembled constitute ourselves prophets of a new order of competence and of courage. This is more than a political campaign; it is a call to arms. Give me your help, not to win votes alone, but to win in this crusade to restore America to its own people.

Although the New Deal frightened many conservatives, Roosevelt worked to allay their fears. He would later say, "No one in the United States believes more firmly than I do in the system of private business, private property, and private profit. It was this administration which saved the system of private profit and free enterprise after it had been dragged to the brink of ruin."

That November, after Hoover conceded defeat, Roosevelt returned to his Manhattan townhouse from campaign headquarters. His mother greeted him. "This is the greatest night of my life," he proclaimed.

Later the same night, his son James came to his bedroom to kiss him good night. James remembers his father looking up at him and saying, "You know, Jimmy, all my life I have been afraid of only one thing—fire. Tonight I think I'm afraid of something else."

"Afraid of what, Pa?"

"I'm afraid that I may not have the strength to do this job." And he asked Jimmy to pray for him.

But when Roosevelt gave his first inaugural address four months later on March 4, 1933, he had found that strength. His speech would become one of the most famous of his entire political career. For with it he galvanized a country frightened by a recent slide even deeper into Depression, with a level of sheer leadership that may never be equaled and by assuring Americans: "The only thing we have to fear is fear itself."

... I AM CERTAIN that my fellow Americans expect that on my induction into the presidency I will address them with a candor and a decision which the present situation of our nation impels. This is preeminently the time to speak the truth, the whole truth, frankly and boldly. Nor need we shrink from honestly facing conditions in our country today. This great nation will endure as it has endured, will revive and will prosper. So, first of all, let me assert my firm belief that the only thing we have to fear is fear itself—nameless, unreasoning, unjustified terror which paralyzes needed efforts to convert retreat into advance. In every dark hour of our national life a leadership of frankness and vigor has met with that understanding and support of the people themselves which is essential to victory. I am convinced that you will again give that support to leadership in these critical days.

In such a spirit on my part and on yours we face our common difficulties. They concern, thank God, only material things. Values have shrunken to fantastic levels; taxes have risen; our ability to pay has fallen; government of all kinds is faced by serious curtailment of income; the means of exchange are frozen in the currents of trade; the withered leaves of industrial enterprise lie on every side; farmers find no markets for their produce; the savings of many years in thousands of families are gone.

More important, a host of unemployed citizens face the grim problem of existence, and an equally great number toil with little return. Only a foolish optimist can deny the dark realities of the moment.

Yet our distress comes from no failure of substance. We are stricken by no plague of locusts. Compared with the perils which our forefathers conquered because they believed and were not afraid, we have still much to be thankful for. Nature still offers her bounty and human efforts have multiplied it. Plenty is at our doorstep, but a generous use of it languishes in the very sight of the supply. Primarily this is because rulers of the exchange of mankind's goods have failed through their own stubbornness and their own incompetence, have admitted their failure, and have abdicated. Practices of the unscrupulous money changers stand indicted in the court of public opinion, rejected by the hearts and minds of men.

True, they have tried, but their efforts have been cast in the pattern of an outworn tradition. Faced by failure of credit they have proposed only the lending of more money. Stripped of the lure of profit by which to induce our people to follow their false leadership, they have resorted to exhortations, pleading tearfully for restored confidence. They know only the rules of a generation of self-seekers. They have no vision, and when there is no vision the people perish.

The money changers have fled from their high seats in the temple of our civilization. We may now restore that temple to the ancient truths. The measure of the restoration lies in the extent to which we apply social values more noble than mere monetary profit.

Happiness lies not in the mere possession of money; it lies in the joy of achievement, in the thrill of creative effort. The joy and moral stimulation of work no longer must be forgotten in the mad chase of evanescent profits. These dark days will be worth all they cost us if they teach us that our true destiny is not to be ministered unto but to minister to ourselves and to our fellow men.

Recognition of the falsity of material wealth as the standard of success goes hand in hand with the abandonment of the false belief that public office and high political position are to be valued only by the standards of pride of place and personal profit; and there must be an end to a conduct in banking and in business which too often has given to a sacred trust the likeness of callous and selfish wrongdoing. Small wonder that confidence languishes, for it thrives only on honesty, on honor, on the sacredness of obligations, on faithful protection, on unselfish performance; without them it cannot live.

Restoration calls, however, not for changes in ethics alone. This nation asks for action, and action now.

Our greatest primary task is to put people to work. This is no unsolvable problem if we face it wisely and courageously. It can be accomplished in part by direct recruiting by the government itself, treating the task as we would treat the emergency of a war, but at the same time, through this employment, accomplishing greatly needed projects to stimulate and reorganize the use of our natural resources.

Hand in hand with this we must frankly recognize the overbalance of population in our industrial centers and, by engaging on a national scale in a redistribution, endeavor to provide a better use of the land for those best fitted for the land. The task can be helped by definite efforts to raise the values of agricultural products and with this the power to purchase the output of our cities. It can be helped by preventing realistically the tragedy of the growing loss through foreclosure of our small homes and our farms. It can be helped by insistence that the federal, state, and local governments act forthwith on the demand that their cost be drastically reduced. It can be helped by the unifying of relief activities which today are often scattered, uneconomical, and unequal. It can be helped by national planning for and supervision of all forms of transportation and of communications and other utilities which have a definitely public character. There are many ways in which it can be helped, but it can never be helped merely by talking about it. We must act and act quickly.

Finally, in our progress toward a resumption of work we require two safeguards against a return of the evils of the old order: there must be a strict supervision of all banking and credits and investments, so that there will be an end to speculation with other people's money; and there must be provision for an adequate but sound currency.

These are the lines of attack. I shall presently urge upon a new Congress, in special session, detailed measures for their fulfillment, and I shall seek the immediate assistance of the several states.

Through this program of action we address ourselves to putting our own national house in order and making income balance outgo. Our international trade relations, though vastly important, are in point of time and necessity secondary to the establishment of a sound national economy. I favor as a practical policy the putting of first things first. I shall spare no effort to restore world trade by international economic readjustment, but the emergency at home cannot wait on that accomplishment.

The basic thought that guides these specific means of national recovery is not narrowly nationalistic. It is the insistence, as a first consideration, upon the interdependence of the various elements in and parts of the United States—a recognition of the old and permanently important manifestation of the American spirit of the pioneer. It is the way to recovery. It is the immediate way. It is the strongest assurance that the recovery will endure.

In the field of world policy I would dedicate this nation to the policy of the Good Neighbor—the neighbor who resolutely respects himself and, because he does so, respects the rights of others—the neighbor who respects his obligations and respects the sanctity of his agreements in and with a world of neighbors.

If I read the temper of our people correctly, we now realize as we have never realized before our interdependence on each other; that we cannot merely take but we must give as well; that if we are to go forward, we must move as a trained and loyal army willing to sacrifice for the good of a common discipline, because without such discipline no progress is made, no leadership becomes effective. We are, I know, ready and willing to submit our lives and property to such discipline, because it makes possible a leadership which aims at a larger good. This I propose to offer, pledging that the larger purposes will bind upon us all as a sacred obligation with a unity of duty hitherto evoked only in time of armed strife.

With this pledge taken, I assume unhesitatingly the

leadership of this great army of our people dedicated to a disciplined attack upon our common problems.

Action in this image and to this end is feasible under the form of government which we have inherited from our ancestors. Our Constitution is so simple and practical that it is possible always to meet extraordinary needs by changes in emphasis and arrangement without loss of essential form. That is why our constitutional system has proved itself the most superbly enduring political mechanism the modern world has produced. It has met every stress of vast expansion of territory, of foreign wars, of bitter internal strife, of world relations.

It is to be hoped that the normal balance of executive and legislative authority may be wholly adequate to meet the unprecedented task before us. But it may be that an unprecedented demand and need for undelayed action may call for temporary departure from that normal balance of public procedure.

I am prepared under my constitutional duty to recommend the measures that a stricken nation in the midst of a stricken world may require. These measures, or such other measures as the Congress may build out of its experience and wisdom, I shall seek, within my constitutional authority, to bring to speedy adoption.

But in the event that the Congress shall fail to take one of these two courses, and in the event that the national emergency is still critical, I shall not evade the clear course of duty that will then confront me. I shall ask the Congress for the one remaining instrument to meet the crisis—broad executive power to wage a war against the emergency, as great as the power that would be given to me if we were in fact invaded by a foreign foe.

For the trust reposed in me I will return the courage and the devotion that befit the time. I can do no less.

We face the arduous days that lie before us in the warm courage of national unity; with the clear consciousness of seeking old and precious moral values; with the clean satisfaction that comes from the stern performance of duty by old and young alike. We aim at the assurance of a rounded and permanent national life.

We do not distrust the future of essential democracy. The people of the United States have not failed. In their need they have registered a mandate that they want direct, vigorous action. They have asked for discipline and direction under leadership. They have made me the present instrument of their wishes. In the spirit of the gift I take it.

In this dedication of a nation we humbly ask the blessing of God. May He protect each and every one of us. May He guide me in the days to come.

In his first and perhaps most famous Fireside Chat, Roosevelt explained to the American people on March 12, 1933, how he intended to resolve the excruciating banking crisis that faced the nation. He also showed his growing mastery of that new communications medium, radio.

I WANT TO talk for a few minutes with the people of the United States about banking—with the comparatively few who understand the mechanics of banking but more particularly with the overwhelming majority who use banks for the making of deposits and the drawing of checks. I want to tell you what has been done in the last few days, why it was done, and what the next steps are going to be. I recognize that the many proclamations from state capitols and from Washington, the legislation, the Treasury regulations, etc., couched for the most part in banking and legal terms, should be explained for the benefit of the average citizen. I owe this in particular because of the fortitude and good temper with which everybody has accepted the inconvenience and hardships of the banking holiday. I know that when you understand what we in Washington have been about I shall continue to have your cooperation as fully as I have had your sympathy and help during the past week.

First of all, let me state the simple fact that when you deposit money in a bank, the bank does not put the money into a safe deposit vault. It invests your money in many different forms of credit—bonds, commercial paper, mortgages, and many other kinds of loans. In other words, the bank puts your money to work to keep the wheels of industry and of agriculture turning around. A comparatively small part of the money you put into the bank is kept in currency—an amount which in normal times is wholly sufficient to cover the cash needs of the average citizen. In other words, the total amount of all the currency in the country is only a small fraction of the total deposits in all of the banks.

What, then, happened during the last few days of February and the first few days of March? Because of undermined confidence on the part of the public, there was a general rush by a large portion of our population to turn bank deposits into currency or gold—a rush so great that the soundest banks could not get enough currency to meet the demand. The reason for this was that on the spur of the moment it was, of course, impossible to sell perfectly sound assets of a bank and convert them into cash except at panic prices far below their real value.

By the afternoon of March 3, scarcely a bank in the country was open to do business. Proclamations temporarily closing them in whole or in part had been issued by the governors in almost all the states.

It was then that I issued the proclamation providing for the nationwide bank holiday, and this was the first step in the government's reconstruction of our financial and economic fabric.

The second step was the legislation promptly and patriotically passed by the Congress confirming my proclamation and broadening my powers so that it became possible in view of the requirement of time to extend the holiday and lift the ban of that holiday gradually. This law also gave authority to develop a program of rehabilitation of our banking facilities. I want to tell our citizens in every part of the nation that the national Congress—Republicans and Democrats alike—showed by this action a devotion to public welfare and a realization of the emergency and the necessity for speed that is difficult to match in our history.

The third stage has been the series of regulations permitting the banks to continue their functions to take care of the distribution of food and household necessities and the payment of payrolls.

This bank holiday, while resulting in many cases in great inconvenience, is affording us the opportunity to supply the currency necessary to meet the situation. No sound bank is a dollar worse off than it was when it closed its doors last Monday. Neither is any bank which may turn out not to be in a position for immediate opening. The new law allows the twelve Federal Reserve Banks to issue additional currency on good assets and thus the banks which reopen will be able to meet every legitimate call. The new currency is being sent out by the Bureau of Engraving and Printing in large volume to every part of the country. It is sound currency because it is backed by actual, good assets.

A question you will ask is this: Why are all the banks not to be reopened at the same time? The answer is simple. Your government does not intend that the history of the past few years shall be repeated. We do not want and will not have another epidemic of bank failures.

As a result, we start tomorrow, Monday, with the opening of banks in the twelve Federal Reserve Bank cities—those banks which on first examination by the Treasury have already been found to be all right. This will be followed on Tuesday by the resumption of all their functions by banks already found to be sound in cities where there are recognized clearinghouses. That means about 250 cities in the United States.

On Wednesday and succeeding days banks in smaller places all through the country will resume business, subject, of course, to the government's physical ability to complete its survey. It is necessary that the reopening of banks be extended over a period in order to permit the banks to make applications for necessary loans, to obtain currency needed to meet their requirements and to enable the government to make common-sense checkups.

Let me make it clear to you that if your bank does not open the first day you are by no means justified in believing that it will not open. A bank that opens on one of the subsequent days is in exactly the same status as the bank that opens tomorrow.

I know that many people are worrying about state

banks not members of the Federal Reserve System. These banks can and will receive assistance from member banks and from the Reconstruction Finance Corporation. These state banks are following the same course as the national banks except that they get their licenses to resume business from the state authorities, and these authorities have been asked by the Secretary of the Treasury to permit their good banks to open up on the same schedule as the national banks. I am confident that the state banking departments will be as careful as the national government in the policy relating to the opening of banks and will follow the same broad policy.

It is possible that when the banks resume a very few people who have not recovered from their fear may again begin withdrawals. Let me make it clear that the banks will take care of all needs—and it is my belief that hoarding during the past week has become an exceedingly unfashionable pastime. It needs no prophet to tell you that when the people find that they can get their money—that they can get it when they want it for all legitimate purposes—the phantom of fear will soon be laid. People will again be glad to have their money where it will be safely taken care of and where they can use it conveniently at any time. I can assure you that it is safer to keep your money in a reopened bank than under the mattress.

The success of our whole great national program depends, of course, upon the cooperation of the public—on its intelligent support and use of a reliable system.

Remember that the essential accomplishment of the new legislation is that it makes it possible for banks more readily to convert their assets into cash than was the case before. More liberal provision has been made for banks to borrow on these assets at the Reserve Banks and more liberal provision has also been made for issuing currency on the security of these good assets. This currency is not fiat currency. It is issued only on adequate security, and every good bank has an abundance of such security.

One more point before I close. There will be, of course, some banks unable to reopen without being reorganized. The new law allows the government to assist in making these reorganizations quickly and effectively and even allows the government to subscribe to at least part of new capital which may be required.

I hope you can see from this elemental recital of what your government is doing that there is nothing complex, or radical, in the process.

We had a bad banking situation. Some of our bankers had shown themselves either incompetent or dishonest in their handling of the people's funds. They had used the money entrusted to them in speculations and unwise loans. This was, of course, not true in the vast majority of our banks, but it was true in enough of them to shock the people for a time into a sense of insecurity and to put them into a frame of mind where they did not differentiate, but seemed to assume that the acts of a comparative few had tainted them all. It was the government's job to straighten out this situation and do it as quickly as possible. And the job is being performed.

I do not promise you that every bank will be reopened or that individual losses will not be suffered, but there will be no losses that possibly could be avoided; and there would have been more and greater losses had we continued to drift. I can even promise you salvation for some at least of the sorely pressed banks. We shall be engaged not merely in reopening sound banks but in the creation of sound banks through reorganization.

It has been wonderful to me to catch the note of confidence from all over the country. I can never be sufficiently grateful to the people for the loyal support they have given me in their acceptance of the judgment that has dictated our course, even though all our processes may not have seemed clear to them.

After all, there is an element in the readjustment of our financial system more important than currency, more important than gold, and that is the confidence of the people. Confidence and courage are the essentials of success in carrying out our plan. You people must have faith; you must not be stampeded by rumors or guesses. Let us unite in banishing fear. We have provided the machinery to restore our financial system; it is up to you to support and make it work.

It is your problem no less than it is mine. Together we cannot fail.

His remarks reassured people who had withdrawn (or, more often, tried to withdraw) their life's savings from banks in cash. He also reassured the banks themselves. The very next day, vast numbers of customers began making deposits again, and within a week, three out of every four banks was open for business.

In his May 7, 1933, Fireside Chat, Roosevelt would follow up on his one from March, as well as explain some new economic policies.

ON A SUNDAY night a week after my inauguration, I used the radio to tell you about the banking crisis and the measures we were taking to meet it. I think that in that way I made clear to the country various facts that might otherwise have been misunderstood and in general provided a means of understanding which did much to restore confidence.

Tonight, eight weeks later, I come for the second time to give you my report; in the same spirit and by the same means to tell you about what we have been doing and what we are planning to do.

Two months ago we were facing serious problems. The country was dying by inches. It was dying because trade and commerce had declined to dangerously low levels; prices for basic commodities were such as to destroy the value of the assets of national institutions such as banks, savings banks, insurance companies, and others. These institutions, because of their great needs, were foreclosing mortgages, calling loans, refusing credit. Thus there was actually in process of destruction the property of millions of people who had borrowed money on that property in terms of dollars which had had an entirely different value from the level of March 1933. That situation in that crisis did not call for any complicated consideration of economic panaceas or fancy plans. We were faced by a condition and not a theory.

There were just two alternatives. The first was to allow the foreclosures to continue, credit to be withheld, and money to go into hiding, thus forcing liquidation and bankruptcy of banks, railroads, and insurance companies and a recapitalizing of all business and all property on a lower level. This alternative meant a continuation of what is loosely called "deflation," the net result of which would have been extraordinary hardships on all property owners and, incidentally, extraordinary hardships on all persons working for wages through an increase in unemployment and a further reduction of the wage scale.

It is easy to see that the result of this course would have not only economic effects of a very serious nature, but social results that might bring incalculable harm. Even before I was inaugurated I came to the conclusion that such a policy was too much to ask the American people to bear. It involved not only a further loss of homes, farms, savings, and wages, but also a loss of spiritual values—the loss of that sense of security for the present and the future so necessary to the peace and contentment of the individual and of his family. When you destroy these things you will find it difficult to establish confidence of any sort in the future. It was clear that mere appeals from Washington for confidence and the mere lending of more money to shaky institutions could not stop this downward course. A prompt program applied as quickly as possible seemed to me not only justified but imperative to our national security. The Congress—and when I say Congress I mean the members of both political parties—fully understood this and gave me generous and intelligent support. The members of Congress realized that the methods of normal times had to be replaced in the emergency by measures which were suited to the serious and pressing requirements of the moment. There was no actual surrender of power, Congress still retained its constitutional authority, and no one has the slightest desire to change the balance of these powers. The function of Congress is to decide what has to be done and to select the appropriate agency to carry out its will. To this policy it has strictly adhered. The only thing that has been happening has been to designate the president as the agency to carry out certain of the purposes of the Congress. This was constitutional and in keeping with the past American tradition.

The legislation which has been passed or is in the process of enactment can properly be considered as part of a well-grounded plan.

First, we are giving opportunity of employment to one-quarter of a million of the unemployed, especially the young men who have dependents, to go into the forestry and flood prevention work. This is a big task because it means feeding, clothing, and caring for nearly twice as many men as we have in the regular army itself. In creating this Civilian Conservation Corps we are killing two birds with one stone. We are clearly enhancing the value of our natural resources, and we are relieving an appreciable amount of actual distress. This great group of men has entered upon its work on a purely voluntary basis; no military training is involved and we are conserving not only our natural resources, but our human resources. One of the great values to this work is the fact that it is direct and requires the intervention of very little machinery.

Second, I have requested the Congress and have secured action upon a proposal to put the great properties owned by our government at Muscle Shoals to work after long years of wasteful inaction, and with this a broad plan for the improvement of a vast area in the Tennessee Valley. It will add to the comfort and happiness of hundreds of thousands of people and the incident benefits will reach the entire nation.

Next, the Congress is about to pass legislation that will greatly ease the mortgage distress among the farmers and the homeowners of the nation, by providing for the easing of the burden of debt now bearing so heavily upon millions of our people.

Our next step in seeking immediate relief is a grant of half a billion dollars to help the states, counties, and municipalities in their duty to care for those who need direct and immediate relief.

The Congress also passed legislation authorizing the sale of beer in such states as desired it. This has already resulted in considerable reemployment and incidentally has provided much-needed tax revenue.

We are planning to ask the Congress for legislation to enable the government to undertake public works, thus stimulating directly and indirectly the employment of many others in well-considered projects.

Further legislation has been taken up which goes much more fundamentally into our economic problems. The Farm Relief Bill seeks by the use of several methods, alone or together, to bring about an increased return to farmers for their major farm products, seeking at the same time to prevent in the days to come disastrous overproduction, which so often in the past has kept farm commodity prices far below a reasonable return. This measure provides wide powers for emergencies. The extent of its use will depend entirely upon what the future has in store.

Well-considered and conservative measures will likewise be proposed which will attempt to give to the industrial workers of the country a more fair wage return, prevent cutthroat competition and unduly long hours for labor, and at the same time encourage each industry to prevent overproduction.

Our railroad bill falls into the same class because it seeks to provide and make certain definite planning by the railroads themselves, with the assistance of the government, to eliminate the duplication and waste that is now resulting in railroad receiverships and continuing operating deficits.

I am certain that the people of this country understand and approve the broad purposes behind these new governmental policies relating to agriculture and industry and transportation. We found ourselves faced with more agricultural products than we could possibly consume ourselves and with surpluses which other nations did not have the cash to buy from us except at prices ruinously low. We found our factories able to turn out more goods than we could possibly consume, and at the same time we were faced with a falling export demand. We found ourselves with more facilities to transport goods and crops than there were goods and crops to be transported. All of this has been caused in large part by a complete lack of planning and a complete failure to understand the danger signals that have been flying ever since the close of the World War. The people of this country have been erroneously encouraged to believe that they could keep on increasing the output of farm and factory indefinitely and that some magician would find ways and means for that increased output to be consumed with reasonable profit to the producer.

Today we have reason to believe that things are a little better than they were two months ago. Industry has picked up, railroads are carrying more freight, farm prices are better, but I am not going to indulge in issuing proclamations of overenthusiastic assurance. We cannot ballyhoo ourselves back to prosperity. I am going to be honest at all times with the people of the country. I do not want the people of this country to take the foolish course of letting this improvement come back on another speculative wave. I do not want the people to believe that because of unjustified optimism we can resume the ruinous practice of increasing our crop output and our factory output in the hope that a kind Providence will find buyers at high prices. Such a course may bring us immediate and false prosperity but it will be the kind of prosperity that will lead us into another tailspin.

It is wholly wrong to call the measures that we have taken government control of farming, industry, and transportation. It is rather a partnership between government and farming and industry and transportation, not partnership in profits, for the profits still go to the citizens, but rather a partnership in planning, and a partnership to see that the plans are carried out.

Let me illustrate with an example. Take the cotton-goods industry. It is probably true that ninety percent of the cotton manufacturers would agree to eliminate starvation wages, would agree to stop long hours of employment, would agree to stop child labor, would agree to prevent an overproduction that would result in unsalable surpluses. But what good is such an agreement if the other ten percent of cotton manufacturers pay starvation wages, require long hours, employ children in their mills, and turn out burdensome surpluses? The unfair ten percent could produce goods so cheaply that the fair ninety percent would be compelled to meet the unfair conditions. Here is where government comes in. Government ought to have the right and will have the right, after surveying and planning for an industry, to prevent, with the assistance of the overwhelming majority of that industry, unfair practices and to enforce this agreement by the authority of government. The so-called antitrust laws were intended to prevent the creation of monopolies and to forbid unreasonable profits to those monopolies. That purpose of the antitrust laws must be continued, but these laws were never intended to encourage the kind of unfair competition that results in long hours, starvation wages, and overproduction.

The same principle applies to farm products and to transportation and every other field of organized private industry.

We are working toward a definite goal, which is to prevent the return of conditions which came very close to destroying what we call modern civilization. The actual accomplishment of our purpose cannot be attained in a day. Our policies are wholly within purposes for which our American constitutional government was established 150 years ago.

I know that the people of this country will understand this and will also understand the spirit in which we are undertaking this policy. I do not deny that we may make mistakes of procedure as we carry out the policy. I have no expectation of making a hit every time I come to bat. What I seek is the highest possible batting average, not only for myself but for the team. Theodore Roosevelt once said to me: "If I can be right 75 percent of the time I shall come up to the fullest measure of my hopes."

Much has been said of late about federal finances and inflation, the gold standard, etc. Let me make the facts very simple and my policy very clear. In the first place, government credit and government currency are really one and the same thing. Behind government currency we have, in addition to the promise to pay, a reserve of gold and a small reserve of silver. In this connection it is worthwhile remembering that in the past the government has agreed to redeem nearly thirty billion of its debts and its currency in gold, and private corporations in this country have agreed to redeem another sixty or seventy billions of securities and mortgages in gold. The government and private corporations were making these agreements when they knew full well that all of the gold in the United States amounted to only between three and four billions and that all of the gold in all of the world amounted to only about eleven billions.

If the holders of these promises to pay started in to demand gold the first-comers would get gold for a few days and they would amount to about one twenty-fifth of the holders of the securities and the currency. The other twenty-four people out of twenty-five, who did not happen to be at the top of the line, would be told politely that there was no more gold left. We have decided to treat all twenty-five in the same way in the interest of justice and the exercise of the constitutional powers of this government. We have placed everyone on the same basis in order that the general good may be preserved.

Nevertheless, gold, and to a partial extent silver, are perfectly good bases for currency, and that is why I decided not to let any of the gold now in the country go out of it.

A series of conditions arose three weeks ago which very readily might have meant, first, a drain on our gold by foreign countries, and second, as a result of that, a flight of American capital, in the form of gold, out of our country. It is not exaggerating the possibility to tell you that such an occurrence might well have taken from us the major part of our gold reserve and resulted in such a further weakening of our government and private credit as to bring on actual panic conditions and the complete stoppage of the wheels of industry.

The administration has the definite objective of raising commodity prices to such an extent that those who have borrowed money will, on the average, be able to repay that money in the same kind of dollar which they borrowed. We do not seek to let them get such a cheap dollar that they will be able to pay back a great deal less than they borrowed. In other words, we seek to correct a wrong and not to create another wrong in the opposite direction. That is why powers are being given to the administration to provide, if necessary, for an enlargement of credit, in order to correct the existing wrong. These powers will be used when, as, and if it may be necessary to accomplish the purpose.

Hand in hand with the domestic situation, which of course is our first concern, is the world situation, and I want to emphasize to you that the domestic situation is inevitably and deeply tied in with the conditions in all of the other nations of the world. In other words, we can get, in all probability, a fair measure of prosperity to return in the United States, but it will not be permanent unless we get a return to prosperity all over the world.

In the conferences which we have held and are holding with the leaders of other nations, we are seeking four great objectives: first, a general reduction of armaments and through this the removal of the fear of invasion and armed attack, and, at the same time, a reduction in armament costs, in order to help in the balancing of government budgets and the reduction of taxation; second, a cutting down of the trade barriers, in order to restart the flow of exchange of crops and goods between nations; third, the setting up of a stabilization of currencies, in order that trade can make contracts ahead; fourth, the reestablishment of friendly relations and greater confidence between all nations.

Our foreign visitors these past three weeks have responded to these purposes in a very helpful way. All of the nations have suffered alike in this great depression. They have all reached the conclusion that each can best be helped by the common action of all. It is in this spirit that our visitors have met with us and discussed our common problems. The international conference that lies before us must succeed. The future of the world demands it and we have each of us pledged ourselves to the best joint efforts to this end.

To you, the people of this country, all of us, the members of the Congress and the members of this administration, owe a profound debt of gratitude. Throughout the depression you have been patient. You have granted us wide powers; you have encouraged us with a widespread approval of our purposes. Every ounce of strength and every resource at our command we have devoted to the end of justifying your confidence. We are encouraged to believe that a wise and sensible beginning has been made. In the present spirit of mutual confidence and mutual encouragement we go forward.

Here is Roosevelt's address to the governing board of the Pan-American Union, on April 12, 1933, in which he fleshed out his concept of the "good neighbor policy"—a policy he would bring up again at the 1940 Democratic convention as war raged in Europe.

I REJOICE IN this opportunity to participate in the celebration of Pan-American Day and to extend on behalf of the people of the United States a fraternal greeting to our sister American republics. The celebration of Pan-American Day in this building, dedicated to international good will and cooperation, exemplifies a unity of thought and purpose among the peoples of this hemisphere. It is a manifestation of the common ideal of mutual helpfulness, sympathetic understanding, and spiritual solidarity.

There is inspiration in the thought that on this day the attention of the citizens of the twenty-one republics of America is focused on the common ties—historical, cultural, economic, and social—which bind them to one another. Common ideals and a community of interest, together with a spirit of cooperation, have led to the realization that the well-being of one nation depends in large measure upon the well-being of its neighbors. It is upon these foundations that Pan-Americanism has been built.

This celebration commemorates a movement based upon the policy of fraternal cooperation. In my inaugural address I stated that I would "dedicate this nation to the policy of the Good Neighbor—the neighbor who resolutely respects himself and, because he does so, respects the rights of others—the neighbor who respects his obligations and respects the sanctity of his agreements in and with a world of neighbors." Never before has the significance of the words "good neighbor" been so manifest in international relations. Never have the need and benefit of neighborly cooperation in every form of human activity been so evident as they are today.

Friendship among nations, as among individuals, calls for constructive efforts to muster the forces of humanity in order that an atmosphere of close understanding and cooperation may be cultivated. It involves mutual obligations and responsibilities, for it is only by sympathetic respect for the rights of others and a scrupulous fulfillment of the corresponding obligations by each member of the community that a true fraternity can be maintained.

The essential qualities of a true Pan-Americanism must be the same as those which constitute a good neighbor, namely, mutual understanding, and, through such understanding, a sympathetic appreciation of the other's point of view. It is only in this manner that we can hope to build up a system of which confidence, friendship, and goodwill are the cornerstones.

In this spirit the people of every republic on our continent are coming to a deep understanding of the fact that the Monroe Doctrine, of which so much has been written and spoken for more than a century, was and is directed at the maintenance of independence by the peoples of the continent. It was aimed and is aimed against the acquisition in any manner of the control of additional territory in this hemisphere by any non-American power.

Hand in hand with this Pan-American doctrine of continental self-defense, the peoples of the American republics understand more clearly, with the passing years, that the independence of each republic must recognize the independence of every other republic. Each one of us must grow by an advancement of civilization and social wellbeing and not by the acquisition of territory at the expense of any neighbor.

In this spirit of mutual understanding and of cooperation on this continent you and I cannot fail to be disturbed by any armed strife between neighbors. I do not hesitate to say to you, the distinguished members of the governing board of the Pan-American Union, that I regard existing conflicts between four of our sister republics as a backward step.

Your Americanism and mine must be a structure built of confidence, cemented by a sympathy which recognizes only equality and fraternity. It finds its source and being in the hearts of men and dwells in the temple of the intellect.

We, all of us, have peculiar problems, and, to speak frankly, the interest of our own citizens must, in each instance, come first. But it is equally true that it is of vital importance to every nation of this continent that the American governments, individually, take, without further delay, such action as may be possible to abolish all unnecessary and artificial barriers and restrictions which now hamper the healthy flow of trade between the peoples of the American republics.

I am glad to deliver this message to you, gentlemen of the Governing Board of the Pan-American Union, for I look upon the Union as the outward expression of the spiritual unity of the Americas. It is to this unity which must be courageous and vital in its element that humanity must look for one of the great stabilizing influences in world affairs.

In closing, may I refer to the ceremony which is to take place a little later in the morning at which the government of Venezuela will present to the Pan-American Union the bust of a great American leader and patriot, Francisco de Miranda. I join with you in this tribute.

On July 24, 1933, Roosevelt held a Fireside Chat to discuss the National Recovery Administration [NRA].

AFTER THE ADJOURNMENT of the historical special session of the Congress five weeks ago I purposely refrained from addressing you for two very good reasons.

First, I think that we all wanted the opportunity of a little quiet thought to examine and assimilate in a mental picture the crowding events of the hundred days which had been devoted to the starting of the wheels of the New Deal.

Secondly, I wanted a few weeks in which to set up the new administrative organization and to see the first fruits of our careful planning.

I think it will interest you if I set forth the fundamentals of this planning for national recovery; and this I am very certain will make it abundantly clear to you that all of the proposals and all of the legislation since the fourth day of March have not been just a collection of haphazard schemes, but rather the orderly component parts of a connected and logical whole.

Long before Inauguration Day I became convinced that individual effort and local effort and even disjointed federal effort had failed and of necessity would fail and, therefore, that a rounded leadership by the federal government had become a necessity both of theory and of fact. Such leadership, however, had its beginning in preserving and strengthening the credit of the United States government, because without that no leadership was a possibility. For years the government had not lived within its income. The immediate task was to bring our regular expenses within our revenues. That has been done.

It may seem inconsistent for a government to cut down its regular expenses and at the same time to borrow and to spend billions for an emergency. But it is not inconsistent because a large portion of the emergency money has been paid out in the form of sound loans which will be repaid to the Treasury over a period of years; and to cover the rest of the emergency money we have imposed taxes to pay the interest and the installments on that part of the debt.

So you will see that we have kept our credit good. We have built a granite foundation in a period of confusion. That foundation of the federal credit stands there broad and sure. It is the base of the whole recovery plan.

Then came the part of the problem that concerned the credit of the individual citizens themselves. You and I know of the banking crisis and of the great danger to the savings of our people. On March 6 every national bank was closed. One month later 90 percent of the deposits in the national banks had been made available to the depositors. Today only about 5 percent of the deposits in national banks are still tied up. The condition relating to state banks, while not quite so good on a percentage basis, is showing a steady reduction in the total of frozen deposits—a result much better than we had expected three months ago.

The problem of the credit of the individual was made more difficult because of another fact. The dollar was a different dollar from the one with which the average debt had been incurred. For this reason large numbers of people were actually losing possession of and title to their farms and homes. All of you know the financial steps which have been taken to correct this inequality. In addition the Home Loan Act, the Farm Loan Act, and the Bankruptcy Act were passed.

It was a vital necessity to restore purchasing power by reducing the debt and interest charges upon our people, but while we were helping people to save their credit it was at the same time absolutely essential to do something about the physical needs of hundreds of thousands who were in dire straits at that very moment. Municipal and state aid were being stretched to the limit. We appropriated half a billion dollars to supplement their efforts and in addition, as you know, we have put three hundred thousand young men into practical and useful work in our forests and to prevent flood and soil erosion. The wages they earn are going in greater part to the support of the nearly one million people who constitute their families.

In this same classification we can properly place the great public works program running to a total of over three billion dollars to be used for highways and ships and flood prevention and inland navigation and thousands of self-sustaining state and municipal improvements. Two points should be made clear in the allotting and administration of these projects: first, we are using the utmost care to choose labor-creating, quick-acting, useful projects, avoiding the smell of the pork barrel; and second, we are hoping that at least half of the money will come back to the government from projects which will pay for themselves over a period of years.

Thus far I have spoken primarily of the foundation stones—the measures that were necessary to reestablish credit and to head people in the opposite direction by preventing distress and providing as much work as possible through governmental agencies. Now I come to the links which will build us a more lasting prosperity. I have said that we cannot attain that in a nation half boom and half broke. If all of our people have work and fair wages and fair profits, they can buy the products of their neighbors, and business is good. But if you take away the wages and the profits of half of them, business is only half as good. It does not help much if the fortunate half is very prosperous; the best way is for everybody to be reasonably prosperous.

For many years the two great barriers to a normal prosperity have been low farm prices and the creeping paralysis of unemployment. These factors have cut the purchasing power of the country in half. I promised action. Congress did its part when it passed the Farm and the Industrial Recovery acts. Today we are putting these two acts to work and they will work if people understand their plain objectives.

First, the Farm Act: it is based on the fact that the purchasing power of nearly half our population depends on adequate prices for farm products. We have been producing more of some crops than we consume or can sell in a depressed world market. The cure is not to produce so much. Without our help the farmers cannot get together and cut production, and the Farm Bill gives them a method of bringing their production down to a reasonable level and of obtaining reasonable prices for their crops. I have clearly stated that this method is in a sense experimental, but so far as we have gone we have reason to believe that it will produce good results.

It is obvious that if we can greatly increase the purchasing power of the tens of millions of our people who make a living from farming and the distribution of farm crops, we shall greatly increase the consumption of those goods which are turned out by industry.

That brings me to the final step—bringing back industry along sound lines.

Last autumn, on several occasions, I expressed my faith that we can make possible by democratic self-discipline in industry general increases in wages and shortening of hours sufficient to enable industry to pay its own workers enough to let those workers buy and use the things that their labor produces. This can be done only if we permit and encourage cooperative action in industry, because it is obvious that without united action a few selfish men in each competitive group will pay starvation wages and insist on long hours of work. Others in that group must either follow suit or close up shop. We have seen the result of action of that kind in the continuing descent into the economic hell of the past four years.

There is a clear way to reverse that process: If all employers in each competitive group agree to pay their workers the same wages—reasonable wages—and require the same hours—reasonable hours—then higher wages and shorter hours will hurt no employer. Moreover, such action is better for the employer than unemployment and low wages, because it makes more buyers for his product. That is the simple idea which is the very heart of the Industrial Recovery Act.

On the basis of this simple principle of everybody doing things together, we are starting out on this nationwide attack on unemployment. It will succeed if our people understand it—in the big industries, in the little shops, in the great cities, and in the small villages. There is nothing complicated about it and there is nothing particularly new in the principle. It goes back to the basic idea of society and of the nation itself that people acting in a group can accomplish things which no individual acting alone could even hope to bring about.

Here is an example. In the Cotton Textile Code and in other agreements already signed, child labor has been abolished. That makes me personally happier than any other one thing with which I have been connected since I came to Washington. In the textile industry—an industry which came to me spontaneously and with a splendid cooperation as soon as the Recovery Act was signed—child labor was an old evil. But no employer acting alone was able to wipe it out. If one employer tried it, or if one state tried it, the costs of operation rose so high that it was impossible to compete with the employers or states which had failed to act. The moment the Recovery Act was passed, this monstrous thing which neither opinion nor law could reach through years of effort went out in a flash. As a British editorial put it, we did more under a code in one day than they in England had been able to do under the common law in eighty-five years of effort. I use this incident, my friends, not to boast of what has already been done but to point the way to you for even greater cooperative efforts this summer and autumn.

We are not going through another winter like the last. I doubt if ever any people so bravely and cheerfully endured a season half so bitter. We cannot ask America to continue to face such needless hardships. It is time for courageous action, and the Recovery Bill gives us the means to conquer unemployment with exactly the same weapon that we have used to strike down child labor.

The proposition is simply this: If all employers will act together to shorten hours and raise wages we can put people back to work. No employer will suffer, because the relative level of competitive cost will advance by the same amount for all. But if any considerable group should lag or shirk, this great opportunity will pass us by and we shall go into another desperate winter. This must not happen.

We have sent out to all employers an agreement which

is the result of weeks of consultation. This agreement checks against the voluntary codes of nearly all the large industries which have already been submitted. This blanket agreement carries the unanimous approval of the three boards which I have appointed to advise in this, boards representing the great leaders in labor, in industry, and in social service. This agreement has already brought a flood of approval from every state, and from so wide a cross-section of the common calling of industry that I know it is fair for all. It is a plan deliberate, reasonable, and just—intended to put into effect at once the most important of the broad principles which are being established, industry by industry, through codes. Naturally, it takes a good deal of organizing and a great many hearings and many months, to get these codes perfected and signed, and we cannot wait for all of them to go through. The blanket agreements, however, which I am sending to every employer will start the wheels turning now, and not six months from now.

There are, of course, men, a few men, who might thwart this great common purpose by seeking selfish advantage. There are adequate penalties in the law, but I am now asking the cooperation that comes from opinion and from conscience. These are the only instruments we shall use in this great summer offensive against unemployment. But we shall use them to the limit to protect the willing from the laggard and to make the plan succeed.

In war, in the gloom of night attack, soldiers wear a bright badge on their shoulders to be sure that comrades do not fire on comrades. On that principle, those who cooperate in this program must know each other at a glance. That is why we have provided a badge of honor for this purpose, a simple design with a legend, "We do our part," and I ask that all those who join with me shall display that badge prominently. It is essential to our purpose.

Already all the great, basic industries have come forward willingly with proposed codes, and in these codes they accept the principles leading to mass reemployment. But, important as is this heartening demonstration, the richest field for results is among the small employers, those whose contribution will be to give new work for from one to ten people. These smaller employers are indeed a vital part of the backbone of the country, and the success of our plan lies largely in their hands.

Already the telegrams and letters are pouring into the White House—messages from employers who ask that their names be placed on this special roll of honor. They represent great corporations and companies, and partnerships and individuals. I ask that even before the dates set in the agreements which we have sent out, the employers of the country who have not already done so—the big fellows and the little fellows—shall at once write or telegraph to me personally at the White House, expressing their intentions of going through with the plan. And it is my purpose to keep posted in the post office of every town, a roll of honor of all those who join with me.

I want to take this occasion to say to the twenty-four governors who are now in conference in San Francisco, that nothing thus far has helped in strengthening this great movement more than their resolutions adopted at the very outset of their meeting, giving this plan their instant and unanimous approval, and pledging to support it in their states.

To the men and women whose lives have been darkened by the fact or the fear of unemployment, I am justified in saying a word of encouragement because the codes and the agreements already approved, or about to be passed upon, prove that the plan does raise wages, and that it does put people back to work. You can look on every employer who adopts the plan as one who is doing his part, and those employers deserve well of everyone who works for a living. It will be clear to you, as it is to me, that while the shirking employer may undersell his competitor, the saving he thus makes is made at the expense of his country's welfare.

While we are making this great common effort there should be no discord and dispute. This is no time to cavil or to question the standard set by this universal agreement. It is time for patience and understanding and cooperation. The workers of this country have rights under this law which cannot be taken from them, and nobody will be permitted to whittle them away but, on the other hand, no aggression is now necessary to attain those rights. The whole country will be united to get them for you. The principle that applies to the employers applies to the workers as well, and I ask you workers to cooperate in the same spirit.

When Andrew Jackson, "Old Hickory," died, someone asked, "Will he go to heaven?" and the answer was, "He will if he wants to." If I am asked whether the American people will pull themselves out of this depression, I answer, "They will if they want to." The essence of the plan is a universal limitation of hours of work per week for any individual by common consent, and a universal payment of wages above a minimum, also by common consent. I cannot guarantee the success of this nationwide plan, but the people of this country can guarantee its success. I have no faith in "cure-alls," but I believe we can greatly influence economic forces. I have no sympathy with the professional economists who insist that things must run their course and that human agencies can have no influence on economic ills. One reason is that I happen to know that professional economists have changed their definition of economic laws every five or ten years for a very long time, but I do have faith, and retain faith, in the strength of the common purpose, and in the strength of unified action taken by the American people.

That is why I am describing to you the simple purposes and the solid foundations upon which our program of recovery is built. That is why I am asking the employers of the nation to sign this common covenant with me—to sign it in the name of patriotism and humanity. That is why I am asking the workers to go along with us in a spirit of understanding and of helpfulness.

On January 3, 1934, Roosevelt delivered his first annual message to Congress. He noted that things had begun to look up for America, but there was some trouble abroad as well as one particular menace at home.

MR. SPEAKER, SENATORS, and Representatives in Congress: I come before you at the opening of the regular session of the Seventy-third Congress not to make requests for special or detailed items of legislation; I come, rather, to counsel with you, who, like myself, have been selected to carry out a mandate of the whole people, in order that without partisanship you and I may cooperate to continue the restoration of our national well-being and, equally important, to build on the ruins of the past a new structure designed better to meet the present problems of modern civilization.

Such a structure includes not only the relations of industry and agriculture and finance to each other but also the effect which all of these three have on our individual citizens and on the whole people as a nation.

Now that we are definitely in the process of recovery, lines have been rightly drawn between those to whom this recovery means a return to old methods—and the number of these people is small—and those for whom recovery means a reform of many old methods, a permanent readjustment of many of our ways of thinking and therefore of many of our social and economic arrangements.

Civilization cannot go back; civilization must not stand still. We have undertaken new methods. It is our task to perfect, to improve, to alter when necessary, but in all cases to go forward. To consolidate what we are doing, to make our economic and social structure capable of dealing with modern life is the joint task of the legislative, the judicial, and the executive branches of the national government.

Without regard to party, the overwhelming majority of our people seek a greater opportunity for humanity to prosper and find happiness. They recognize that human welfare has not increased and does not increase through mere materialism and luxury, but that it does progress through integrity, unselfishness, responsibility, and justice.

In the past few months, as a result of our action, we have demanded of many citizens that they surrender certain licenses to do as they please in their business relationships; but we have asked this in exchange for the protection which the state can give against exploitation by their fellow men or by combinations of their fellow men.

I congratulate this Congress upon the courage, the earnestness, and the efficiency with which you met the crisis at the special session. It was your fine understanding of the national problem that furnished the example which the country has so splendidly followed. I venture to say that the task confronting the First Congress of 1789 was no greater than your own.

I shall not attempt to set forth either the many phases of the crisis which we experienced last March, or the many measures which you and I undertook during the special session that we might initiate recovery and reform.

It is sufficient that I should speak in broad terms of the results of our common counsel.

The credit of the government has been fortified by drastic reduction in the cost of its permanent agencies through the Economy Act.

With the twofold purpose of strengthening the whole financial structure and of arriving eventually at a medium of exchange which over the years will have less variable purchasing and debt-paying power for our people than that of the past, I have used the authority granted me to purchase all American-produced gold and silver and to buy additional gold in the world markets. Careful investigation and constant study prove that in the matter of foreign exchange rates certain of our sister nations find themselves so handicapped by internal and other conditions that they feel unable at this time to enter into stabilization discussion based on permanent and worldwide objectives.

The overwhelming majority of the banks, both national and state, which reopened last spring, are in sound condition and have been brought within the protection of federal insurance. In the case of those banks which were not permitted to reopen, nearly $600 million of frozen deposits are being restored to the depositors through the assistance of the national government.

We have made great strides toward the objectives of the National Industrial Recovery Act, for not only have several millions of our unemployed been restored to work, but industry is organizing itself with a greater understanding that reasonable profits can be earned while at the same time protection can be assured to guarantee to labor adequate pay and proper conditions of work. Child labor is abolished. Uniform standards of hours and wages apply today to 95 percent of industrial employment within the field of the National Industrial Recovery Act. We seek the definite end of preventing combinations in furtherance of monopoly and in restraint of trade, while at the same time we seek to prevent ruinous rivalries within industrial groups which in many cases resemble the gang wars of the underworld and in which the real victim in every case is the public itself. . . .

You recognized last spring that the most serious part of the debt burden affected those who stood in danger of losing their farms and their homes. I am glad to tell you that refinancing in both of these cases is proceeding with good success and in all probability within the financial limits set by the Congress.

But agriculture had suffered from more than its debts. Actual experience with the operation of the Agricultural Adjustment Act leads to my belief that thus far the experiment of seeking a balance between production and consumption is succeeding and has made progress entirely in line with reasonable expectations toward the restoration of farm prices to parity. I continue in my conviction that industrial progress and prosperity can only be attained by bringing the purchasing power of that portion of our population which in one form or another is dependent upon agriculture up to a level which will restore a proper balance between every section of the country and between every form of work.

In this field, through carefully planned flood control, power development, and land-use policies in the Tennessee Valley and in other great watersheds, we are seeking the elimination of waste, the removal of poor lands from agriculture, and the encouragement of small local industries, thus furthering this principle of a better balanced national life. . . .

I cannot, unfortunately, present to you a picture of complete optimism regarding world affairs.

The delegation representing the United States has worked in close cooperation with the other American republics assembled at Montevideo to make that conference an outstanding success. We have, I hope, made it clear to our neighbors that we seek with them future avoidance of territorial expansion and of interference by one nation in the internal affairs of another. Furthermore, all of us are seeking the restoration of commerce in ways which will preclude the building up of large favorable trade balances by any one nation at the expense of trade debits on the part of other nations.

In other parts of the world, however, fear of immediate or future aggression and with it the spending of vast sums on armament and the continued building up of defensive trade barriers prevent any great progress in peace or trade agreements. I have made it clear that the United States cannot take part in political arrangements in Europe but that we stand ready to cooperate at any time in practicable measures on a world basis looking to immediate reduction of armaments and the lowering of the barriers against commerce.

I expect to report to you later in regard to debts owed the government and people of this country by the governments and peoples of other countries. Several nations, acknowledging the debt, have paid in small part; other nations have failed to pay. One nation—Finland—has paid the installments due this country in full.

Returning to home problems, we have been shocked by many notorious examples of injuries done our citizens by persons or groups who have been living off their neighbors by the use of methods either unethical or criminal.

In the first category—a field which does not involve violations of the letter of our laws—practices have been brought to light which have shocked those who believe that we were in the past generation raising the ethical standards of business. They call for stringent preventive or regulatory measures. I am speaking of those individuals who have evaded the spirit and purpose of our tax laws, of those high officials of banks or corporations who have grown rich at the expense of their stockholders or the public, of those reckless speculators with their own or other people's money whose operations have injured the values of the farmers' crops and the savings of the poor.

In the other category, crimes of organized banditry, cold-blooded shooting, lynching, and kidnapping have threatened our security.

These violations of ethics and these violations of law call on the strong arm of government for their immediate suppression; they call also on the country for an aroused public opinion.

The adoption of the Twenty-first Amendment should give material aid to the elimination of those new forms of crime which came from the illegal traffic in liquor.

I shall continue to regard it as my duty to use whatever means may be necessary to supplement state, local, and private agencies for the relief of suffering caused by unemployment. With respect to this question, I have recognized the dangers inherent in the direct giving of relief and have sought the means to provide not mere relief, but the opportunity for useful and remunerative work. We shall, in the process of recovery, seek to move as rapidly as possible from direct relief to publicly supported work and from that to the rapid restoration of private employment.

It is to the eternal credit of the American people that this tremendous readjustment of our national life is being accomplished peacefully, without serious dislocation, with only a minimum of injustice, and with a great, willing spirit of cooperation throughout the country.

Disorder is not an American habit. Self-help and self-control are the essence of the American tradition—not of necessity the form of that tradition, but its spirit. The program itself comes from the American people. . . .

We have plowed the furrow and planted the good seed; the hard beginning is over. If we would reap the full harvest, we must cultivate the soil where this good seed is sprouting and the plant is reaching up to mature growth.

A final personal word. I know that each of you will appreciate that I am speaking no mere politeness when I assure you how much I value the fine relationship that we have shared during these months of hard and incessant work. Out of these friendly contacts we are, fortunately, building a strong and permanent tie between the legislative and executive branches of the government. The letter of the Constitution wisely declared a separation, but the impulse of common purpose declares a union. In this spirit we join once more in serving the American people.

By his second annual message to Congress on January 4, 1935, Roosevelt's complex political philosophies had almost completely evolved. He echoed his explanation for the Temporary Emergency Relief Administration, which he had established as the governor of New York and which by 1932, was helping nearly one family out of ten in New York State: "It is clear to me that it is the duty of those who have benefited by our industrial and economic system to come to the front in such a grave emergency and assist in relieving those who under the same system are the losers and sufferers."

He echoed as well his thoughts upon signing the Social Security Act: "We can never insure one hundred percent of the population against one hundred percent of the hazards and vicissitudes of life, but we have tried to frame a law which will give some measure of protection to the average citizen and his family against the loss of a job and against a poverty-ridden old age."

As both beliefs and many others that he'd stoked America's recovery, were woven together and expanded, Roosevelt revealed himself to be a thoroughly mature and worldly statesman.

MR. SPEAKER, MEMBERS of the Senate and of the House of Representatives: The Constitution wisely provides that the chief executive shall report to the Congress on the state of the Union, for through you, the chosen legislative representatives, our citizens everywhere may fairly judge the progress of our governing. I am confident that today, in the light of the events of the past two years, you do not consider it merely a trite phrase when I tell you that I am truly glad to greet you and that I look forward to common counsel, to useful cooperation, and to genuine friendships between us.

We have undertaken a new order of things; yet we progress to it under the framework and in the spirit and intent of the American Constitution. We have proceeded throughout the nation a measurable distance on the road toward this new order. Materially, I can report to you substantial benefits to our agricultural population, increased industrial activity, and profits to our merchants. Of equal moment, there is evident a restoration of that spirit of confidence and faith which marks the American character. Let him who for speculative profit or partisan purpose, without just warrant, would seek to disturb or dispel this assurance take heed before he assumes responsibility for any act which slows our onward steps.

Throughout the world, change is the order of the day. In every nation, economic problems, long in the making, have brought crises of many kinds for which the masters of old practice and theory were unprepared. In most nations social justice, no longer a distant ideal, has become a definite goal, and ancient governments are beginning to heed the call.

Thus, the American people do not stand alone in the world in their desire for change. We seek it through tested liberal traditions, through processes which retain all of the deep essentials of that republican form of representative government first given to a troubled world by the United States. . . .

We find our population suffering from old inequalities, little changed by past sporadic remedies. In spite of our efforts and in spite of our talk, we have not weeded out the overprivileged and we have not effectively lifted up the underprivileged. Both of these manifestations of injustice have retarded happiness. No wise man has any intention of destroying what is known as the profit motive; because by the profit motive we mean the right by work to earn a decent livelihood for ourselves and for our families.

We have, however, a clear mandate from the people, that Americans must forswear that conception of the acquisition of wealth which, through excessive profits, creates undue private power over private affairs and, to our misfortune, over public affairs as well. In building toward this end we do not destroy ambition, nor do we seek to divide our wealth into equal shares on stated occasions. We continue to recognize the greater ability of some to earn more than others. But we do assert that the ambition of the individual to obtain for him and his a proper security, a reasonable leisure, and a decent living throughout life is an ambition to be preferred to the appetite for great wealth and great power.

I recall to your attention my message to the Congress last June in which I said: "Among our objectives I place the security of the men, women, and children of the nation first." That remains our first and continuing task; and in a very real sense every major legislative enactment of this Congress should be a component part of it.

In defining immediate factors which enter into our quest, I have spoken to the Congress and the people of three great divisions:

1. The security of a livelihood through the better use of the national resources of the land in which we live.

2. The security against the major hazards and vicissitudes of life.

3. The security of decent homes.

I am now ready to submit to the Congress a broad program designed ultimately to establish all three of these factors of security—a program which because of many lost years will take many future years to fulfill.

A study of our national resources, more comprehensive than any previously made, shows the vast amount of necessary and practicable work which needs to be done for the development and preservation of our natural wealth for the enjoyment and advantage of our people in generations to come. The sound use of land and water is far more comprehensive than the mere planting of trees, building of dams, distributing of electricity, or retirement of sub-marginal land. It recognizes that stranded populations, either in the country or the city, cannot have security under the conditions that now surround them.

To this end we are ready to begin to meet this problem—the intelligent care of population throughout our nation, in accordance with an intelligent distribution of the means of livelihood for that population. A definite program for putting people to work, of which I shall speak in a moment, is a component part of this greater program of security of livelihood through the better use of our national resources.

Closely related to the broad problem of livelihood is that of security against the major hazards of life. Here also, a comprehensive survey of what has been attempted or accomplished in many nations and in many states proves to me that the time has come for action by the national government. I shall send to you, in a few days, definite recommendations based on these studies. These recommendations will cover the broad subjects of unemployment insurance and old-age insurance, of benefits for children, for mothers, for the handicapped, for maternity care, and for other aspects of dependency and illness where a beginning can now be made.

The third factor—better homes for our people—has also been the subject of experimentation and study. Here, too, the first practical steps can be made through the proposals which I shall suggest in relation to giving work to the unemployed. . . .

The first objectives of emergency legislation of 1933 were to relieve destitution, to make it possible for industry to operate in a more rational and orderly fashion, and to put behind industrial recovery the impulse of large expenditures in government undertakings. The purpose of the National Industrial Recovery Act to provide work for more people succeeded in a substantial manner within the first few months of its life, and the act has continued to maintain employment gains and greatly improved working conditions in industry. . . .

More than two billions of dollars have also been expended in direct relief to the destitute. Local agencies of necessity determined the recipients of this form of relief. With inevitable exceptions the funds were spent by them with reasonable efficiency and as a result actual want of food and clothing in the great majority of cases has been overcome.

But the stark fact before us is that great numbers still remain unemployed.

A large proportion of these unemployed and their dependents have been forced on the relief rolls. The burden on the federal government has grown with great rapidity. We have here a human as well as an economic problem. When humane considerations are concerned, Americans give them precedence. The lessons of history, confirmed by the evidence immediately before me, show conclusively that continued dependence upon relief induces a spiritual and moral disintegration fundamentally destructive to the national fiber. To dole out relief in this way is to administer a narcotic, a subtle destroyer of the human spirit. It is inimical to the dictates of sound policy. It is in violation of the traditions of America. Work must be found for able-bodied but destitute workers.

The federal government must and shall quit this business of relief.

I am not willing that the vitality of our people be further sapped by the giving of cash, of market baskets, of a few hours of weekly work cutting grass, raking leaves, or picking up papers in the public parks. We must preserve not only the bodies of the unemployed from destitution but also their self-respect, their self-reliance and courage and determination. This decision brings me to the problem of what the government should do with approximately five million unemployed now on the relief rolls.

About one million and a half of these belong to the group which in the past was dependent upon local welfare efforts. Most of them are unable for one reason or another to maintain themselves independently—for the most part, through no fault of their own. Such people, in the days before the Great Depression, were cared for by local efforts—by states, by counties, by towns, by cities, by churches, and by private welfare agencies. It is my thought that in the future they must be cared for as they were before. . . .

The security legislation which I shall propose to the Congress will, I am confident, be of assistance to local effort in the care of this type of cases. Local responsibility can and will be resumed, for, after all, common sense tells us that the wealth necessary for this task existed and still exists in the local community, and the dictates of sound administration require that this responsibility be in the first instance a local one.

There are, however, an additional three and one-half million employable people who are on relief. With them the problem is different and the responsibility is different. This group was the victim of a nationwide depression caused by conditions which were not local but national. The federal government is the only governmental agency with sufficient power and credit to meet this situation. We have assumed this task and we shall not shrink from it in the future. It is a duty dictated by every intelligent consideration of national policy to ask you to make it possible for the United States to give employment to all of these three and one-half million employable people now on relief, pending their absorption in a rising tide of private employment.

It is my thought that with the exception of certain of the normal public building operations of the government, all emergency public works shall be united in a single new and greatly enlarged plan.

With the establishment of this new system we can supersede the Federal Emergency Relief Administration with a coordinated authority which will be charged with the orderly liquidation of our present relief activities and the substitution of a national chart for the giving of work.

This new program of emergency public employment should be governed by a number of practical principles.

1. All work undertaken should be useful—not just for a day, or a year, but useful in the sense that it affords permanent improvement in living conditions or that it creates future new wealth for the nation.

2. Compensation on emergency public projects should be in the form of security payments which should be larger than the amount now received as a relief dole, but at the same time not so large as to encourage the rejection of opportunities for private employment or the leaving of private employment to engage in government work.

3. Projects should be undertaken on which a large percentage of direct labor can be used.

4. Preference should be given to those projects which will be self-liquidating in the sense that there is a reasonable expectation that the government will get its money back at some future time.

5. The projects undertaken should be selected and planned so as to compete as little as possible with private enterprises. This suggests that if it were not for the necessity of giving useful work to the unemployed now on relief, these projects in most instances would not now be undertaken.

6. The planning of projects would seek to assure work during the coming fiscal year to the individuals now on relief, or until such time as private employment is available. In order to make adjustment to increasing private employment, work should be planned with a view to tapering it off in proportion to the speed with which the

emergency workers are offered positions with private employers.

7. Effort should be made to locate projects where they will serve the greatest unemployment needs as shown by present relief rolls, and the broad program of the National Resources Board should be freely used for guidance in selection. Our ultimate objective being the enrichment of human lives, the government has the primary duty to use its emergency expenditures as much as possible to serve those who cannot secure the advantages of private capital. . . .

The work itself will cover a wide field including clearance of slums, which for adequate reasons cannot be undertaken by private capital; in rural housing of several kinds, where, again, private capital is unable to function; in rural electrification; in the reforestation of the great watersheds of the nation; in an intensified program to prevent soil erosion and to reclaim blighted areas; in improving existing road systems and in constructing national highways designed to handle modern traffic; in the elimination of grade crossings; in the extension and enlargement of the successful work of the Civilian Conservation Corps; in nonfederal works, mostly self-liquidating and highly useful to local divisions of government; and on many other projects which the nation needs and cannot afford to neglect. . . .

We have already begun to feel the bracing effect upon our economic system of a restored agriculture. The hundreds of millions of additional income that farmers are receiving are finding their way into the channels of trade. The farmers' share of the national income is slowly rising. The economic facts justify the widespread opinion of those engaged in agriculture that our provisions for maintaining a balanced production give at this time the most adequate remedy for an old and vexing problem. For the present, and especially in view of abnormal world conditions, agricultural adjustment with certain necessary improvements in methods should continue. . . .

I cannot with candor tell you that general international relationships outside the borders of the United States are improved. On the surface of things many old jealousies are resurrected, old passions aroused; new strivings for armament and power, in more than one land, rear their ugly heads. I hope that calm counsel and constructive leadership will provide the steadying influence and the time necessary for the coming of new and more practical forms of representative government throughout the world wherein privilege and power will occupy a lesser place and world welfare a greater.

I believe, however, that our own peaceful and neighborly attitude toward other nations is coming to be understood and appreciated. The maintenance of international peace is a matter in which we are deeply and unselfishly concerned. Evidence of our persistent and undeniable desire to prevent armed conflict has recently been more than once afforded. . . .

The ledger of the past year shows many more gains than losses. Let us not forget that, in addition to saving millions from utter destitution, child labor has been for the moment outlawed, thousands of homes saved to their owners, and most important of all, the morale of the nation has been restored. Viewing the year 1934 as a whole, you and I can agree that we have a generous measure of reasons for giving thanks.

It is not empty optimism that moves me to a strong hope in the coming year. We can, if we will, make 1935 a genuine period of good feeling, sustained by a sense of purposeful progress. Beyond the material recovery, I sense a spiritual recovery as well. The people of America are turning as never before to those permanent values that are not limited to the physical objectives of life. There are growing signs of this on every hand. In the face of these spiritual impulses we are sensible of the Divine Providence to which nations turn now, as always, for guidance and fostering care.

During his Fireside Chat of April 28, 1935, Roosevelt expanded upon his message to Congress by explaining the Works Relief Program to the American people.

SINCE MY ANNUAL Message to the Congress on January fourth last, I have not addressed the general public over the air. In the many weeks since that time the Congress has devoted itself to the arduous task of formulating legislation necessary to the country's welfare. It has made and is making distinct progress.

Before I come to any of the specific measures, however, I want to leave in your minds one clear fact. The administration and the Congress are not proceeding in any haphazard fashion in this task of government. Each of our steps has a definite relationship to every other step. The job of creating a program for the nation's welfare is, in some respects, like the building of a ship. At different points on the coast where I often visit they build great seagoing ships. When one of these ships is under construction and the steel frames have been set in the keel, it is difficult for a person who does not know ships to tell how it will finally look when it is sailing the high seas.

It may seem confused to some, but out of the multitude of detailed parts that go into the making of the structure, the creation of a useful instrument for man ultimately comes. It is that way with the making of a national policy. The objective of the nation has greatly changed in three years. Before that time individual self-interest and group selfishness were paramount in public thinking. The general good was at a discount.

Three years of hard thinking have changed the picture. More and more people, because of clearer thinking and a better understanding, are considering the whole rather than a mere part relating to one section, or to one crop, or to one industry, or to an individual private occupation. That is a tremendous gain for the principles of democracy. The overwhelming majority of people in this country know how to sift the wheat from the chaff in what they hear and what they read. They know that the process of the constructive rebuilding of America cannot be done in a day or a year, but that it is being done in spite of the few who seek to confuse them and to profit by their confusion. Americans as a whole are feeling a lot better—a lot more cheerful than for many, many years.

The most difficult place in the world to get a clear and open perspective of the country as a whole is Washington. I am reminded sometimes of what President Wilson once said: "So many people come to Washington who know things that are not so, and so few people who know what the people of the United States are thinking about." That is why I occasionally leave this scene of action for a few days to go fishing or back home to Hyde Park so that I can have a chance to think quietly about the country as a whole. "To get away from the trees," as they say, "and to look at the whole forest." This duty of seeing the country in a long-range perspective is one which, in a very special manner, attaches to this office to which you have chosen me. Did you ever stop to think that there are, after all, only two positions in the nation that are filled by the vote of all of the voters—the president and the vice president? That makes it particularly necessary for the vice president and for me to conceive of our duty toward the entire country. Tonight, therefore, I speak to and of the American people as a whole.

My most immediate concern is in carrying out the purposes of the great work program just enacted by the Congress. Its first objective is to put men and women now on the relief rolls to work and, incidentally, to assist materially in our already unmistakable march toward recovery. I shall not confuse my discussion by a multitude of figures. So many figures are quoted to prove so many things. Sometimes it depends upon what paper you read and what broadcast you hear. Therefore, let us keep our minds on two or three simple, essential facts in connection with this problem of unemployment. It is true that while business and industry are definitely better, our relief rolls are still too large. However, for the first time in five years the relief rolls have declined instead of increased during the winter months. They are still declining. The simple fact is that many millions more people have private work today than two years ago today or one year ago today, and every day that passes offers more chances to work for those who want to work. In spite of the fact that unemployment remains a serious problem here as in every other nation, we have come to recognize the possibility and the necessity of certain helpful remedial measures.

These measures are of two kinds. The first is to make provisions intended to relieve, to minimize, and to prevent future unemployment; the second is to establish the practical means to help those who are unemployed in this present emergency. Our Social Security legislation is an attempt to answer the first of these questions; our Works Relief program, the second.

The program for Social Security now pending before the Congress is a necessary part of the future unemployment policy of the government. While our present and projected expenditures for work relief are wholly within the reasonable limits of our national credit resources, it is obvious that we cannot continue to create governmental deficits for that purpose year after year. We must begin now to make provision for the future. That is why our Social Security program is an important part of the complete picture. It proposes, by means of old-age pensions, to help those who have reached the age of retirement to give up their jobs and thus give to the younger generation greater opportunities for work and to give to all a feeling of security as they look toward old age.

The unemployment-insurance part of the legislation will not only help to guard the individual in future periods of layoff against dependence upon relief, but it will, by sustaining purchasing power, cushion the shock of economic distress. Another helpful feature of unemployment insurance is the incentive it will give to employers to plan more carefully in order that unemployment may be prevented by the stabilizing of employment itself.

Provisions for Social Security, however, are protections for the future. Our responsibility for the immediate necessities of the unemployed has been met by the Congress through the most comprehensive work plan in the history of the nation. Our problem is to put to work three and one-half million employable persons now on the relief rolls. It is a problem quite as much for private industry as for the government.

We are losing no time getting the government's vast Works Relief program under way and we have every reason to believe that it should be in full swing by autumn. In directing it, I shall recognize six fundamental principles:

1. The projects should be useful.

2. Projects shall be of a nature that a considerable proportion of the money spent will go into wages for labor.

3. Projects will be sought which promise ultimate return to the federal Treasury of a considerable proportion of the costs.

4. Funds allotted for each project should be actually and promptly spent and not held over until later years.

5. In all cases projects must be of a character to give employment to those on the relief rolls.

6. Projects will be allocated to localities or relief areas in relation to the number of workers on relief rolls in those areas.

I next want to make it clear exactly how we shall direct the work.

1. I have set up a Division of Applications and Information to which all proposals for the expenditure of money must go for preliminary study and consideration.

2. After the Division of Applications and Information has sifted these projects, they will be sent to an Allotment Division composed of representatives of the more important governmental agencies charged with carrying on work relief projects. The group will also include representatives of cities, and of labor, farming, banking, and indus-

try. This Allotment Division will consider all of the recommendations submitted to it and such projects as they approve will be next submitted to the president who under the act is required to make final allocations.

3. The next step will be to notify the proper government agency in whose field the project falls, and also to notify another agency which I am creating—a Progress Division. This division will have the duty of coordinating the purchase of materials and supplies and of making certain that people who are employed will be taken from the relief rolls. It will also have the responsibility of determining work payments in various localities, of making full use of existing employment services, and of assisting people engaged in relief work to move as rapidly as possible back into private employment when such employment is available. Moreover, this division will be charged with keeping projects moving on schedule.

4. I have felt it to be essentially wise and prudent to avoid, so far as possible, the creation of new governmental machinery for supervising this work. The national government now has at least sixty different agencies with the staff and the experience and the competence necessary to carry on the two hundred and fifty or three hundred kinds of work that will be undertaken. These agencies, therefore, will simply be doing on a somewhat enlarged scale the same sort of things that they have been doing. This will make certain that the largest possible portion of the funds allotted will be spent for actually creating new work and not for building up expensive overhead organizations here in Washington.

For many months preparations have been under way. The allotment of funds for desirable projects has already begun. The key men for the major responsibilities of this great task already have been selected. I well realize that the country is expecting before this year is out to see the "dirt fly," as they say, in carrying on the work, and I assure my fellow citizens that no energy will be spared in using these funds effectively to make a major attack upon the problem of unemployment.

Our responsibility is to all of the people in this country. This is a great national crusade to destroy enforced idleness which is an enemy of the human spirit generated by this depression. Our attack upon these enemies must be without stint and without discrimination. No sectional, no political distinctions can be permitted.

It must, however, be recognized that when an enterprise of this character is extended over more than three thousand counties throughout the nation, there may be occasional instances of inefficiency, bad management, or misuse of funds. When cases of this kind occur, there will be those, of course, who will try to tell you that the exceptional failure is characteristic of the entire endeavor. It should be remembered that in every big job there are some imperfections. There are chiselers in every walk of life, there are those in every industry who are guilty of unfair practices; every profession has its black sheep, but long experience in government has taught me that the exceptional instances of wrongdoing in government are probably less numerous than in almost every other line of endeavor. The most effective means of preventing such evils in this Works Relief program will be the eternal vigilance of the American people themselves. I call upon my fellow citizens everywhere to cooperate with me in making this the most efficient and the cleanest example of public enterprise the world has ever seen.

It is time to provide a smashing answer for those cynical men who say that a democracy cannot be honest and efficient. If you will help, this can be done. I, therefore, hope you will watch the work in every corner of this nation. Feel free to criticize. Tell me of instances where work can be done better, or where improper practices prevail. Neither you nor I want criticism conceived in a purely fault-finding or partisan spirit, but I am jealous of the right of every citizen to call to the attention of his or her government examples of how the public money can be more effectively spent for the benefit of the American people.

I now come, my friends, to a part of the remaining business before the Congress. It has under consideration many measures which provide for the rounding out of the program of economic and social reconstruction with which we have been concerned for two years. I can mention only a few of them tonight, but I do not want my mention of specific measures to be interpreted as lack of interest in or disapproval of many other important proposals that are pending.

The National Industrial Recovery Act expires on the sixteenth of June. After careful consideration, I have asked the Congress to extend the life of this useful agency of government. As we have proceeded with the administration of this act, we have found from time to time more and more useful ways of promoting its purposes. No reasonable person wants to abandon our present gains—we must continue to protect children, to enforce minimum wages, to prevent excessive hours, to safeguard, define, and enforce collective bargaining, and, while retaining fair competition, to eliminate, so far as humanly possible, the kinds of unfair practices by selfish minorities which unfortunately did more than anything else to bring about the recent collapse of industries.

There is likewise pending before the Congress legislation to provide for the elimination of unnecessary holding companies in the public utility field.

I consider this legislation a positive recovery measure. Power production in this country is virtually back to the 1929 peak. The operating companies in the gas and electric utility field are by and large in good condition. But under holding company domination the utility industry has long been hopelessly at war within itself and with public sentiment. By far the greater part of the general decline in utility securities had occurred before I was inaugurated. The absentee management of unnecessary holding company control has lost touch with, and has lost the sympathy of, the communities it pretends to serve. Even more significantly it has given the country as a whole an uneasy apprehension of overconcentrated economic power.

A business that loses the confidence of its customers and the good will of the public cannot long continue to be a good risk for the investor. This legislation will serve the investor by ending the conditions which have caused that lack of confidence and good will. It will put the public utility operating industry on a sound basis for the future, both in its public relations and in its internal relations.

This legislation will not only in the long run result in providing lower electric and gas rates to the consumer but it will protect the actual value and earning power of properties now owned by thousands of investors who have little protection under the old laws against what used to be called frenzied finance. It will not destroy values.

Not only business recovery, but the general economic recovery of the nation will be greatly stimulated by the enactment of legislation designed to improve the status of our transportation agencies. There is need for legislation for the regulation of interstate transportation by buses and trucks, for the regulation of transportation by water, for the strengthening of our Merchant Marine and Air Transport, for the strengthening of the Interstate Commerce Commission to enable it to carry out a rounded conception of the national transportation system in which the benefits of private ownership are retained while the public stake in these important services is protected by the public's government.

Finally, the reestablishment of public confidence in the banks of the nation is one of the most hopeful results of our efforts as a nation to reestablish public confidence in private banking. We all know that private banking actually exists by virtue of the permission of and regulation by the people as a whole, speaking through their government. Wise public policy, however, requires not only that banking be safe but that its resources be most fully utilized in the economic life of the country. To this end it was decided more than twenty years ago that the government should assume the responsibility of providing a means by which the credit of the nation might be controlled, not by a few private banking institutions, but by a body with public prestige and authority. The answer to this demand was the Federal Reserve System. Twenty years of experience with this system have justified the efforts made to create it, but these twenty years have shown by experience definite possibilities for improvement. Certain proposals made to amend the Federal Reserve Act deserve prompt and favorable action by the Congress. They are a minimum of wise readjustments of our Federal Reserve System in the light of past experience and present needs.

These measures I have mentioned are, in large part, the program which under my constitutional duty I have recommended to the Congress. They are essential factors in a rounded program for national recovery. They contemplate the enrichment of our national life by a sound and rational ordering of its various elements and wise provisions for the protection of the weak against the strong.

Never since my inauguration in March 1933 have I felt so unmistakably the atmosphere of recovery. But it is more than the recovery of the material basis of our individual lives. It is the recovery of confidence in our democratic processes and institutions. We have survived all of the arduous burdens and the threatening dangers of a great economic calamity. We have in the darkest moments of our national trials retained our faith in our own ability to master our destiny. Fear is vanishing and confidence is growing on every side, faith is being renewed in the vast possibilities of human beings to improve their material and spiritual status through the instrumentality of the democratic form of government. That faith is receiving its just reward. For that we can be thankful to the God who watches over America.

At the 1936 Democratic convention on June 27 in Philadelphia, accepting his party's nomination, he made one of his most famous speeches, which included the resonant phrase "a rendez-

vous with destiny." In it Roosevelt slammed the "economic royalists" and "privileged princes" who continually countered his policies.

SENATOR ROBINSON, MEMBERS of the Democratic Convention, my friends: Here, and in every community throughout the land, we are met at a time of great moment to the future of the nation. It is an occasion to be dedicated to the simple and sincere expression of an attitude toward problems, the determination of which will profoundly affect America.

I come not only as a leader of a party, not only as a candidate for high office, but as one upon whom many critical hours have imposed and still impose a grave responsibility.

For the sympathy, help and confidence with which Americans have sustained me in my task I am grateful. For their loyalty I salute the members of our great party, in and out of political life in every part of the Union. I salute those of other parties, especially those in the Congress of the United States who on so many occasions have put partisanship aside. I thank the governors of the several states, their legislatures, their state and local officials who participated unselfishly and regardless of party in our efforts to achieve recovery and destroy abuses. Above all I thank the millions of Americans who have borne disaster bravely and have dared to smile through the storm.

America will not forget these recent years, will not forget that the rescue was not a mere party task. It was the concern of all of us. In our strength we rose together, rallied our energies together, applied the old rules of common sense, and together survived.

In those days we feared fear. That was why we fought fear. And today, my friends, we have won against the most dangerous of our foes. We have conquered fear.

But I cannot, with candor, tell you that all is well with the world. Clouds of suspicion, tides of ill-will and intolerance gather darkly in many places. In our own land we enjoy indeed a fullness of life greater than that of most nations. But the rush of modern civilization itself has raised for us new difficulties, new problems which must be solved if we are to preserve to the United States the political and economic freedom for which Washington and Jefferson planned and fought.

Philadelphia is a good city in which to write American history. This is fitting ground on which to reaffirm the faith of our fathers; to pledge ourselves to restore to the people a wider freedom; to give to 1936 as the founders gave to 1776—an American way of life.

That very word, *freedom,* in itself and of necessity, suggests freedom from some restraining power. In 1776 we sought freedom from the tyranny of a political autocracy from the eighteenth-century royalists who held special privileges from the crown. It was to perpetuate their privilege that they governed without the consent of the governed; that they denied the right of free assembly and free speech; that they restricted the worship of God; that they put the average man's property and the average man's life in pawn to the mercenaries of dynastic power; that they regimented the people.

And so it was to win freedom from the tyranny of political autocracy that the American Revolution was fought. That victory gave the business of governing into the hands of the average man, who won the right with his neighbors to make and order his own destiny through his own government. Political tyranny was wiped out at Philadelphia on July 4, 1776.

Since that struggle, however, man's inventive genius released new forces in our land which reordered the lives of our people. The age of machinery, of railroads; of steam and electricity; the telegraph and the radio; mass production, mass distribution—all of these combined to bring forward a new civilization and with it a new problem for those who sought to remain free.

For out of this modern civilization economic royalists carved new dynasties. New kingdoms were built upon concentration of control over material things. Through new uses of corporations, banks and securities, new machinery of industry and agriculture, of labor and capital all undreamed of by the Fathers the whole structure of modern life was impressed into this royal service.

There was no place among this royalty for our many thousands of small-businessmen and merchants who sought to make a worthy use of the American system of initiative and profit. They were no more free than the worker or the farmer. Even honest and progressive-minded men of wealth, aware of their obligation to their generation, could never know just where they fitted into this dynastic scheme of things.

It was natural and perhaps human that the privileged princes of these new economic dynasties, thirsting for power, reached out for control over government itself. They created a new despotism and wrapped it in the robes of legal sanction. In its service new mercenaries sought to regiment the people, their labor, and their property. And as a result the average man once more confronts the problem that faced the Minute Man.

The hours men and women worked, the wages they

received, the conditions of their labor—these had passed beyond the control of the people, and were imposed by this new industrial dictatorship. The savings of the average family, the capital of the small-businessman, the investments set aside for old age—other people's money—these were tools which the new economic royalty used to dig itself in.

Those who tilled the soil no longer reaped the rewards which were their right. The small measure of their gains was decreed by men in distant cities.

Throughout the nation, opportunity was limited by monopoly. Individual initiative was crushed in the cogs of a great machine. The field open for free business was more and more restricted. Private enterprise, indeed, became too private. It became privileged enterprise, not free enterprise.

For too many of us the political equality we once had won was meaningless in the face of economic inequality. A small group had concentrated into their own hands an almost complete control over other people's property, other people's money, other people's labor, other people's lives. For too many of us life was no longer free; liberty no longer real; men could no longer follow the pursuit of happiness.

Against economic tyranny such as this, the American citizen could appeal only to the organized power of government. The collapse of 1929 showed up the despotism for what it was. The election of 1932 was the people's mandate to end it. Under that mandate it is being ended.

The royalists of the economic order have conceded that political freedom was the business of the government, but they have maintained that economic slavery was nobody's business. They granted that the government could protect the citizen in his right to vote, but they denied that the government could do anything to protect the citizen in his right to work and his right to live.

Today we stand committed to the proposition that freedom is no half-and-half affair. If the average citizen is guaranteed equal opportunity in the polling place, he must have equal opportunity in the market place.

These economic royalists complain that we seek to overthrow the institutions of America. What they really complain of is that we seek to take away their power. Our allegiance to American institutions requires the overthrow of this kind of power. In vain they seek to hide behind the flag and the Constitution. In their blindness they forget what the flag and the Constitution stand for. Now, as always, they stand for democracy, not tyranny; for freedom, not subjection; and against a dictatorship by mob rule and the overprivileged alike.

The defeats and victories of these years have given to us as a people a new understanding of our government and of ourselves. Never since the early days of the New England town meeting have the affairs of government been so widely discussed and so clearly appreciated. It has been brought home to us that the only effective guide for the safety of this most worldly of worlds, the greatest guide of all, is moral principle.

We do not see faith, hope, and charity as unattainable ideals, but we use them as stout supports of a nation fighting the fight for freedom in a modern civilization.

Faith—in the soundness of democracy in the midst of dictatorships.

Hope—renewed because we know so well the progress we have made.

Charity—in the true spirit of that grand old word. For

charity literally translated from the original means love, the love that understands, that does not merely share the wealth of the giver, but in true sympathy and wisdom helps men to help themselves.

We seek not merely to make government a mechanical implement, but to give it the vibrant personal character that is the very embodiment of human charity.

We are poor indeed if this nation cannot afford to lift from every recess of American life the dread fear of the unemployed that they are not needed in the world. We cannot afford to accumulate a deficit in the books of human fortitude.

In the place of the palace of privilege we seek to build a temple out of faith and hope and charity.

It is a sobering thing, my friends, to be a servant of this great cause. We try in our daily work to remember that the cause belongs not to us, but to the people. The standard is not in the hands of you and me alone. It is carried by America. We seek daily to profit from experience, to learn to do better as our task proceeds.

Governments can err, presidents do make mistakes, but the immortal Dante tells us that Divine justice weighs the sins of the cold-blooded and the sins of the warm-hearted in different scales.

Better the occasional faults of a government that lives in a spirit of charity than the consistent omissions of a government frozen in the ice of its own indifference.

There is a mysterious cycle in human events. To some generations much is given. Of other generations much is expected. This generation of Americans has a rendezvous with destiny.

In this world of ours in other lands, there are some people, who, in times past, have lived and fought for freedom, and seem to have grown too weary to carry on the fight. They have sold their heritage of freedom for the illusion of a living. They have yielded their democracy.

I believe in my heart that only our success can stir their ancient hope. They begin to know that here in America we are waging a great and successful war. It is not alone a war against want and destitution and economic demoralization. It is more than that; it is a war for the survival of democracy. We are fighting to save a great and precious form of government for ourselves and for the world.

I accept the commission you have tendered me. I join with you. I am enlisted for the duration of the war.

At Madison Square Garden, FDR said, "I should like to have it said of my first administration that in it the forces of selfishness and lust for power have met their match. And I should like to have it said of my second administration that in it these forces have met their master." During his highly dramatic second inaugural address, delivered in pouring rain on January 20, 1937, Roosevelt would echo this sentiment as he showed the American people the facts they had to face.

WHEN FOUR YEARS ago we met to inaugurate a president, the Republic, single-minded in anxiety, stood in spirit here. We dedicated ourselves to the fulfillment of a vision to speed the time when there would be for all the people that security and peace essential to the pursuit of happiness. We of the Republic pledged ourselves to drive from the temple of our ancient faith those who had profaned it; to end by action, tireless and unafraid, the stagnation and despair of that day. We did those first things first.

Our covenant with ourselves did not stop there. Instinctively we recognized a deeper need—the need to find through government the instrument of our united purpose to solve for the individual the ever-rising problems of a complex civilization. Repeated attempts at their solution without the aid of government had left us baffled and bewildered. For, without that aid, we had been unable to create those moral controls over the services of science which are necessary to make science a useful servant instead of a ruthless master of mankind. To do this we knew that we must find practical controls over blind economic forces and blindly selfish men.

We of the Republic sensed the truth that democratic government has innate capacity to protect its people against disasters once considered inevitable, to solve problems once considered unsolvable. We would not admit that we could not find a way to master economic epidemics just as, after centuries of fatalistic suffering, we had found a way to master epidemics of disease. We refused to leave the problems of our common welfare to be solved by the winds of chance and the hurricanes of disaster.

In this we Americans were discovering no wholly new truth; we were writing a new chapter in our book of self-government.

This year marks the hundred and fiftieth anniversary of the Constitutional Convention which made us a nation. At that Convention our forefathers found the way out of the chaos which followed the Revolutionary War; they created a strong government with powers of united action sufficient then and now to solve problems utterly beyond individual or local solution. A century and a half ago they established the federal government in order to promote the general welfare and secure the blessings of liberty to the American people.

Today we invoke those same powers of government to achieve the same objectives.

Four years of new experience have not belied our historic instinct. They hold out the clear hope that government within communities, government within the separate states, and government of the United States can do the things the times require, without yielding its democracy. Our tasks in the last four years did not force democracy to take a holiday.

Nearly all of us recognize that as intricacies of human relationships increase, so power to govern them also must increase power to stop evil; power to do good. The essential democracy of our nation and the safety of our people depend not upon the absence of power, but upon lodging it with those whom the people can change or continue at stated intervals through an honest and free system of elections. The Constitution of 1787 did not make our democracy impotent.

In fact, in these last four years, we have made the exercise of all power more democratic; for we have begun to bring private autocratic powers into their proper subordination to the public's government. The legend that they were invincible, above and beyond the processes of a democracy, has been shattered. They have been challenged and beaten.

Our progress out of the depression is obvious. But that is not all that you and I mean by the new order of things. Our pledge was not merely to do a patchwork job with secondhand materials. By using the new materials of social justice we have undertaken to erect on the old foundations a more enduring structure for the better use of future generations.

In that purpose we have been helped by achievements of mind and spirit. Old truths have been relearned; untruths have been unlearned. We have always known that heedless self-interest was bad morals; we know now that it is bad economics. Out of the collapse of a prosperity whose builders boasted their practicality has come the conviction that in the long run economic morality pays. We are beginning to wipe out the line that divides the practical from the ideal; and in so doing we are fashioning an instrument of unimagined power for the establishment of a morally better world.

This new understanding undermines the old admiration of worldly success as such. We are beginning to abandon our tolerance of the abuse of power by those who betray for profit the elementary decencies of life.

In this process evil things formerly accepted will not be so easily condoned. Hard-headedness will not so easily excuse hard-heartedness. We are moving toward an era of good feeling. But we realize that there can be no era of good feeling save among men of goodwill.

For these reasons I am justified in believing that the greatest change we have witnessed has been the change in the moral climate of America.

Among men of goodwill, science and democracy together offer an ever-richer life and ever-larger satisfaction to the individual. With this change in our moral climate and our rediscovered ability to improve our economic order, we have set our feet upon the road of enduring progress.

Shall we pause now and turn our back upon the road that lies ahead? Shall we call this the promised land? Or, shall we continue on our way? For "each age is a dream that is dying, or one that is coming to birth."

Many voices are heard as we face a great decision. Comfort says, "Tarry a while." Opportunism says, "This is a good spot." Timidity asks, "How difficult is the road ahead?"

True, we have come far from the days of stagnation and despair. Vitality has been preserved. Courage and confidence have been restored. Mental and moral horizons have been extended.

But our present gains were won under the pressure of more than ordinary circumstance. Advance became imperative under the goad of fear and suffering. The times were on the side of progress.

To hold to progress today, however, is more difficult. Dulled conscience, irresponsibility, and ruthless self-interest already reappear. Such symptoms of prosperity may become portents of disaster! Prosperity already tests the persistence of our progressive purpose.

Let us ask again: Have we reached the goal of our vision of that fourth day of March 1933? Have we found our happy valley?

I see a great nation, upon a great continent, blessed with a great wealth of natural resources. Its hundred and thirty million people are at peace among themselves; they are making their country a good neighbor among the nations. I see a United States which can demonstrate that, under democratic methods of government, national wealth can be translated into a spreading volume of human comforts hitherto unknown, and the lowest standard of living can be raised far above the level of mere subsistence.

But here is the challenge to our democracy: in this

nation I see tens of millions of its citizens—a substantial part of its whole population who at this very moment are denied the greater part of what the very lowest standards of today call the necessities of life.

I see millions of families trying to live on incomes so meager that the pall of family disaster hangs over them day by day.

I see millions whose daily lives in city and on farm continue under conditions labeled indecent by a so-called polite society half a century ago.

I see millions denied education, recreation, and the opportunity to better their lot and the lot of their children.

I see millions lacking the means to buy the products of farm and factory and by their poverty denying work and productiveness to many other millions.

I see one-third of a nation ill-housed, ill-clad, ill-nourished.

It is not in despair that I paint you that picture. I paint it for you in hope because the nation, seeing and understanding the injustice in it, proposes to paint it out. We are determined to make every American citizen the subject of his country's interest and concern; and we will never regard any faithful, law-abiding group within our borders as superfluous. The test of our progress is not whether we add more to the abundance of those who have much; it is whether we provide enough for those who have too little.

If I know aught of the spirit and purpose of our nation, we will not listen to Comfort, Opportunism, and Timidity. We will carry on.

Overwhelmingly, we of the Republic are men and women of goodwill; men and women who have more than warm hearts of dedication; men and women who have cool heads and willing hands of practical purpose as well. They will insist that every agency of popular government use effective instruments to carry out their will.

Government is competent when all who compose it work as trustees for the whole people. It can make constant progress when it keeps abreast of all the facts. It can obtain justified support and legitimate criticism when the people receive true information of all that government does.

If I know aught of the will of our people, they will demand that these conditions of effective government shall be created and maintained. They will demand a nation uncorrupted by cancers of injustice and, therefore, strong among the nations in its example of the will to peace.

Today we reconsecrate our country to long-cherished ideals in a suddenly changed civilization. In every land there are always at work forces that drive men apart and forces that draw men together. In our personal ambitions we are individualists. But in our seeking for economic and political progress as a nation, we all go up, or else we all go down, as one people.

To maintain a democracy of effort requires a vast amount of patience in dealing with differing methods, a vast amount of humility. But out of the confusion of many voices rises an understanding of dominant public need. Then political leadership can voice common ideals, and aid in their realization.

In taking again the oath of office as president of the United States, I assume the solemn obligation of leading the American people forward along the road over which they have chosen to advance.

While this duty rests upon me I shall do my utmost

to speak their purpose and to do their will, seeking Divine guidance to help us each and every one to give light to them that sit in darkness and to guide our feet into the way of peace.

Unhappy that the Supreme Court had declared several of his programs unconstitutional during his first administration, Roosevelt then made one of the few serious mistakes in his career. He tried to "pack" the Supreme Court. Here's how he presented his idea to the American people during his Fireside Chat of March 9, 1937.

. . . . IT IS THE clear intention of our public policy to provide for a constant flow of new and younger blood into the judiciary. Normally every president appoints a large number of district and circuit judges and a few members of the Supreme Court. Until my first term practically every president of the United States had appointed at least one member of the Supreme Court. President Taft appointed five members and named a Chief Justice; President Wilson, three; President Harding, four, including a Chief Justice; President Coolidge, one; President Hoover, three, including a Chief Justice.

Such a succession of appointments should have provided a Court well-balanced as to age. But chance and the disinclination of individuals to leave the Supreme bench have now given us a Court in which five justices will be over seventy-five years of age before next June and one over seventy. Thus a sound public policy has been defeated.

I now propose that we establish by law an assurance against any such ill-balanced Court in the future. I propose that hereafter, when a judge reaches the age of seventy, a new and younger judge shall be added to the Court automatically. In this way I propose to enforce a sound public policy by law instead of leaving the composition of our federal courts, including the highest, to be determined by chance or the personal decision of individuals. . . .

This proposal of mine will not infringe in the slightest upon the civil or religious liberties so dear to every American.

My record as governor and as president proves my devotion to those liberties. You who know me can have no fear that I would tolerate the destruction by any branch of government of any part of our heritage of freedom. . . .

During the past half century the balance of power between the three great branches of the federal government has been tipped out of balance by the courts in direct contradiction of the high purposes of the framers of the Constitution. It is my purpose to restore that balance. You who know me will accept my solemn assurance that in a world in which democracy is under attack, I seek to make American democracy succeed. You and I will do our part.

Despite his efforts in this and other Fireside Chats, as well as in press conferences to build public support for his plan, Congress refused to allow it. Although Roosevelt eventually gave up, the Court's decisions took a much more liberal turn in the ensuing years as several conservative justices stepped down and were replaced by justices more friendly to Roosevelt's administration.

In his speeches and letters throughout the mid-1930s, he continually expressed his intentions not to draw America into another war. "I hate war. I have passed unnumbered hours, I shall pass unnumbered hours, thinking and planning how war may be kept from this nation."

But by 1937, he clearly perceived a "reign of terror and international lawlessness" in the growth of the Axis powers of Germany, Italy, and Japan. "Innocent peoples . . . innocent nations are being cruelly sacrificed to a greed for power." So on October 5, during a speech intended merely to dedicate a bridge in Chicago, FDR took the opportunity to sound the warning bell against isolationism.

I AM GLAD to come once again to Chicago and especially to have the opportunity of taking part in the dedication of this important project of civic betterment.

On my trip across the continent and back, I have been shown many evidences of the result of common-sense cooperation between municipalities and the federal government, and I have been greeted by tens of thousands of Americans who have told me in every look and word that their material and spiritual well-being has made great strides forward in the past few years.

And yet, as I have seen with my own eyes the prosperous farms, the thriving factories, and the busy railroads, as I have seen the happiness and security and peace which covers our wide land, almost inevitably I have been compelled to contrast our peace with very different scenes being enacted in other parts of the world.

It is because the people of the United States under modern conditions must, for the sake of their own future, give thought to the rest of the world, that I, as the responsible executive head of the nation, have chosen this great inland city and this gala occasion to speak to you on a subject of definite national importance.

The political situation in the world, which of late has been growing progressively worse, is such as to cause grave concern and anxiety to all the peoples and nations who wish to live in peace and amity with their neighbors.

Some fifteen years ago the hopes of mankind for a continuing era of international peace were raised to great heights when more than sixty nations solemnly pledged themselves not to resort to arms in furtherance of their national aims and policies. The high aspirations expressed in the Briand-Kellogg peace pact and the hopes for peace thus raised have of late given way to a haunting fear of calamity. The present reign of terror and international lawlessness began a few years ago.

It began through unjustified interference in the internal affairs of other nations or the invasion of alien territory in violation of treaties, and has now reached a stage where the very foundations of civilization are seriously threatened. The landmarks and traditions which have marked the progress of civilization toward a condition of law, order, and justice are being wiped away.

Without a declaration of war and without warning or justification of any kind, civilians, including vast numbers of women and children, are being ruthlessly murdered with bombs from the air. In times of so-called peace, ships are being attacked and sunk by submarines without cause or notice. Nations are fomenting and taking sides in civil warfare in nations that have never done them any harm. Nations claiming freedom for themselves deny it to others.

Innocent peoples, innocent nations, are being cruelly

sacrificed to a greed for power and supremacy which is devoid of all sense of justice and humane considerations. . . .

The peace-loving nations must make a concerted effort in opposition to those violations of treaties and those ignorings of humane instincts which today are creating a state of international anarchy and instability from which there is no escape through mere isolation or neutrality.

Those who cherish their freedom, and recognize and respect the equal right of their neighbors to be free and live in peace, must work together for the triumph of law and moral principles in order that peace, justice, and confidence may prevail in the world. There must be a return to a belief in the pledged word, in the value of a signed treaty. There must be recognition of the fact that national morality is as vital as private morality. . . .

There is a solidarity and interdependence about the modern world, both technically and morally, which makes it impossible for any nation completely to isolate itself from economic and political upheavals in the rest of the world, especially when such upheavals appear to be spreading and not declining. There can be no stability or peace either within nations or between nations except under laws and moral standards adhered to by all. International anarchy destroys every foundation for peace. It jeopardizes either the immediate or the future security of every nation, large or small. It is, therefore, a matter of vital interest and concern to the people of the United States that the sanctity of international treaties and the maintenance of international morality be restored. . . .

It seems to be unfortunately true that the epidemic of world lawlessness is spreading.

When an epidemic of physical disease starts to spread,

the community approves and joins in a quarantine of the patients in order to protect the health of the community against the spread of the disease.

It is my determination to pursue a policy of peace and to adopt every practicable measure to avoid involvement in war. It ought to be inconceivable that in this modern era, and in the face of experience, any nation could be so foolish and ruthless as to run the risk of plunging the whole world into war by invading and violating, in contravention of solemn treaties, the territory of other nations that have done them no real harm and are too weak to protect themselves adequately. Yet the peace of the world and the welfare and security of every nation, including our own, is today being threatened by that very thing.

No nation which refuses to exercise forbearance and to respect the freedom and rights of others can long remain strong and retain the confidence and respect of other nations. No nation ever loses its dignity or its good standing by conciliating its differences, and by exercising great patience with, and consideration for, the rights of other nations.

War is a contagion, whether it be declared or undeclared. It can engulf states and peoples remote from the original scene of hostilities. We are determined to keep out of war, yet we cannot insure ourselves against the disastrous effects of war and the dangers of involvement. We are adopting such measures as will minimize our risk of involvement, but we cannot have complete protection in a world of disorder in which confidence and security have broken down.

If civilization is to survive, the principles of the Prince of Peace must be restored. Trust between nations must be revived.

Most important of all, the will for peace on the part of peace-loving nations must express itself to the end that nations that may be tempted to violate their agreements and the rights of others will desist from such a course. There must be positive endeavors to preserve peace.

America hates war. America hopes for peace. Therefore, America actively engages in the search for peace.

In his sixth annual message to Congress, on January 4, 1939, Roosevelt not only made state-of-the-union remarks, he also gave Congress a keen sense of his philosophies and increasingly sophisticated world view as the specter of war loomed over the world.

A WAR WHICH threatened to envelop the world in flames has been averted; but it has become increasingly clear that world peace is not assured.

All about us rage undeclared wars military and economic. All about us grow more deadly armaments military and economic. All about us are threats of new aggression military and economic.

Storms from abroad directly challenge three institutions indispensable to Americans, now as always. The first is religion. It is the source of the other two—democracy and international good faith.

Religion, by teaching man his relationship to God, gives the individual a sense of his own dignity and teaches him to respect himself by respecting his neighbors.

Democracy, the practice of self-government, is a covenant among free men to respect the rights and liberties of their fellows.

International good faith, a sister of democracy, springs from the will of civilized nations of men to respect the rights and liberties of other nations of men.

In a modern civilization, all three—religion, democracy, and international good faith—complement and support each other.

Where freedom of religion has been attacked, the attack has come from sources opposed to democracy. Where democracy has been overthrown, the spirit of free worship has disappeared. And where religion and democracy have vanished, good faith and reason in international affairs have given way to strident ambition and brute force.

We know what might happen to us of the United States if the new philosophies of force were to encompass the other continents and invade our own. We, no more than other nations, can ill afford to be surrounded by the enemies of our faith and our humanity. Fortunate it is, therefore, that in this Western Hemisphere we have, under a common ideal of democratic government, a rich diversity of resources and of peoples functioning together in mutual respect and peace.

That hemisphere, that peace, and that ideal we propose to do our share in protecting against storms from any quarter. Our people and our resources are pledged to secure that protection. From that determination, no American flinches.

In our foreign relations we have learned from the past what not to do. From new wars we have learned what we must do.

We have learned that effective timing of defense, and the distant points from which attacks may be launched are completely different from what they were twenty years ago.

We have learned that survival cannot be guaranteed by arming after the attack begins for there is new range and speed to offense.

We have learned that long before any overt military act, aggression begins with preliminaries of propaganda, subsidized penetration, the loosening of ties of goodwill, the stirring of prejudice, and the incitement to disunion.

We have learned that God-fearing democracies of the world which observe the sanctity of treaties and good faith in their dealings with other nations cannot safely be indifferent to international lawlessness anywhere. They cannot forever let pass, without effective protest, acts of aggression against sister nations—acts which automatically undermine all of us.

Obviously they must proceed along practical, peaceful lines. But the mere fact that we rightly decline to intervene with arms to prevent acts of aggression does not mean that we must act as if there were no aggression at all. Words may be futile, but war is not the only means of commanding a decent respect for the opinions of mankind. There are many methods short of war, but stronger and more effective than mere words, of bringing home to aggressor governments the aggregate sentiments of our own people.

At the very least, we can and should avoid any action, or any lack of action, which will encourage, assist, or build up an aggressor. We have learned that when we deliberately try to legislate neutrality, our neutrality laws may operate unevenly and unfairly—may actually give aid to an aggressor and deny it to the victim. The instinct of self-preservation should warn us that we ought not to let that happen anymore.

And we have learned something else—the old, old lesson that probability of attack is mightily decreased by the assurance of an ever-ready defense. Since 1931, nearly eight years ago, world events of thunderous import have moved with lightning speed. During these eight years, many of our people clung to the hope that the innate decency of mankind would protect the unprepared who showed their innate trust in mankind. Today we are all wiser—and sadder.

Under modern conditions what we mean by "adequate defense"—a policy subscribed to by all of us—must be divided into three elements. First, we must have armed forces and defenses strong enough to ward off sudden attack against strategic positions and key facilities essential to ensure sustained resistance and ultimate victory. Secondly, we must have the organization and location of those key facilities so that they may be immediately utilized and rapidly expanded to meet all needs without danger of serious interruption by enemy attack. . . .

In meeting the troubles of the world we must meet them as one people with a unity born of the fact that for generations those who have come to our shores, representing many kindreds and tongues, have been welded by common opportunity into a united patriotism. If another form of government can present a united front in its attack on a democracy, the attack must and will be met by a united democracy. Such a democracy can and must exist in the United States.

A dictatorship may command the full strength of a regimented nation. But the united strength of a democratic nation can be mustered only when its people, educated by modern standards to know what is going on and where they are going, have conviction that they are receiving as large a share of opportunity for development, as large a share of material success and of human dignity, as they have a right to receive.

Our nation's program of social and economic reform is therefore a part of defense, as basic as armaments themselves.

Against the background of events in Europe, in Africa, and in Asia during these recent years, the pattern of what we have accomplished since 1933 appears in even clearer focus.

For the first time we have moved upon deep-seated problems affecting our national strength and have forged national instruments adequate to meet them.

Consider what the seemingly piecemeal struggles of these six years add up to in terms of realistic national preparedness.

We are conserving and developing natural resources—land, water power, forests.

We are trying to provide necessary food, shelter, and medical care for the health of our population.

We are putting agriculture—our system of food and fiber supply—on a sounder basis.

We are strengthening the weakest spot in our system of industrial supply—its long-smoldering labor difficulties.

We have cleaned up our credit system so that depositor and investor alike may more readily and willingly make their capital available for peace or war.

We are giving to our youth new opportunities for work and education.

We have sustained the morale of all the population by the dignified recognition of our obligations to the aged, the helpless, and the needy.

Above all, we have made the American people conscious of their interrelationship and their interdependence. They sense a common destiny and a common need of each other. Differences of occupation, geography, race, and religion no longer obscure the nation's fundamental unity in thought and in action.

We have our difficulties, true—but we are a wiser and a tougher nation than we were in 1929, or in 1932.

Never have there been six years of such far-flung internal preparedness in our history. And this has been done without any dictator's power to command, without conscription of labor or confiscation of capital, without concentration camps and without a scratch on freedom of speech, freedom of the press, or the rest of the Bill of Rights. . . .

We have now passed the period of internal conflict in the launching of our program of social reform. Our full energies may now be released to invigorate the processes of recovery in order to preserve our reforms, and to give every man and woman who wants to work a real job at a living wage.

But time is of paramount importance. The deadline of danger from within and from without is not within our control. The hourglass may be in the hands of other nations. Our own hourglass tells us that we are off on a race to make democracy work, so that we may be efficient in peace and therefore secure in national defense.

Events abroad have made it increasingly clear to the American people that dangers within are less to be feared than dangers from without. If, therefore, a solution of this problem of idle men and idle capital is the price of preserving our liberty, no formless, selfish fears can stand in the way.

Once I prophesied that this generation of Americans

had a rendezvous with destiny. That prophecy comes true. To us much is given; more is expected.

This generation will "nobly save or meanly lose the last best hope of earth. . . ." The way is plain, peaceful, generous, just—a way which if followed the world will forever applaud and God must forever bless.

As it grew clear that Hitler had his sights on Poland, on April 14, 1939, Roosevelt delivered a public message to the Führer and Mussolini.

YOU REALIZE, I am sure, that throughout the world hundreds of millions of human beings are living today in constant fear of a new war or even a series of wars.

The existence of this fear and the possibility of such a conflict are of definite concern to the people of the United States for whom I speak, as they must also be to the peoples of the other nations of the entire Western Hemisphere. All of them know that any major war, even if it were to be confined to other continents, must bear heavily on them during its continuance and also for generations to come.

Because of the fact that after the acute tension in which the world has been living during the past few weeks there would seem to be at least a momentary relaxation because no troops are at this moment on the march this may be an opportune moment for me to send you this message.

On a previous occasion I have addressed you in behalf of the settlement of political, economic, and social problems by peaceful methods and without resort to arms.

But the tide of events seems to have reverted to the

threat of arms. If such threats continue, it seems inevitable that much of the world must become involved in common ruin. All the world—victor nations, vanquished nations, and neutral nations will suffer. I refuse to believe that the world is, of necessity, such a prisoner of destiny. On the contrary, it is clear that the leaders of great nations have it in their power to liberate their peoples from the disaster that impends. It is equally clear that in their own minds and in their own hearts the peoples themselves desire that their fears be ended.

It is, however, unfortunately necessary to take cognizance of recent facts.

Three nations in Europe and one in Africa have seen their independent existence terminated. A vast territory in another independent nation of the Far East has been occupied by a neighboring state. Reports, which we trust are not true, insist that further acts of aggression are contemplated against still other independent nations. Plainly the world is moving toward the moment when this situation must end in catastrophe unless a more rational way of guiding events is found.

You have repeatedly asserted that you and the German people have no desire for war. If this is true there need be no war.

Nothing can persuade the peoples of the earth that any governing power has any right or need to inflict the consequences of war on its own or any other people save in the cause of self-evident home defense.

In making this statement we as Americans speak not through selfishness or fear or weakness. If we speak now it is with the voice of strength and with friendship for mankind. It is still clear to me that international problems can be solved at the council table.

It is therefore no answer to the plea for peaceful discussion for one side to plead that unless they receive assurances beforehand that the verdict will be theirs, they will not lay aside their arms. In conference rooms, as in courts, it is necessary that both sides enter upon the discussion in good faith, assuming that substantial justice will accrue to both; and it is customary and necessary that they leave their arms outside the room where they confer.

I am convinced that the cause of world peace would be greatly advanced if the nations of the world were to obtain a frank statement relating to the present and future policy of governments.

Because the United States, as one of the nations of the Western Hemisphere, is not involved in the immediate controversies which have arisen in Europe, I trust that you may be willing to make such a statement of policy to me as head of a nation far removed from Europe in order that I, acting only with the responsibility and obligation of a friendly intermediary, may communicate such declaration to other nations now apprehensive as to the course which the policy of your government may take.

We recognize complex world problems which affect all humanity but we know that study and discussion of them must be held in an atmosphere of peace. Such an atmosphere of peace cannot exist if negotiations are overshadowed by the threat of force or by the fear of war.

I think you will not misunderstand the spirit of frankness in which I send you this message. Heads of great governments in this hour are literally responsible for the fate of humanity in the coming years. They cannot fail to hear the prayers of their peoples to be protected from the foreseeable chaos of war. History will hold them accountable for the lives and the happiness of all even unto the least.

I hope that your answer will make it possible for humanity to lose fear and regain security for many years to come.

Hitler would not, however, be distracted from his chosen course; he didn't even bother to reply to the president's entreaties. So as Germany prepared to invade Poland, Roosevelt again addressed him publicly on August 24, 1939.

IN THE MESSAGE which I sent to you on April 14 last I stated that it appeared to me that the leaders of great nations had it in their power to liberate their peoples from the disaster that impended, but that unless the effort were immediately made with goodwill on all sides to find a peaceful and constructive solution of existing controversies, the crisis which the world was confronting must end in catastrophe. Today that catastrophe appears to be very near at hand indeed.

To the message which I sent to you last April I have received no reply, but because of my confident belief that the cause of world peace which is the cause of humanity itself rises above all other considerations, I am again addressing myself to you with the hope that the war which impends and the consequent disaster to all peoples everywhere may yet be averted.

I therefore urge with all earnestness and I am likewise urging the president of the Republic of Poland that the governments of Germany and of Poland agree by common accord to refrain from any positive act of hostility for a reasonable and stipulated period, and that they agree likewise by common accord to solve the controversies which have arisen between them by one of the three following methods: first, by direct negotiation; second, by submission of these controversies to an impartial arbitration in which they can both have confidence; or, third, that they agree to the solution of these controversies through the procedure of conciliation, selecting as conciliator or moderator a national of one of the traditionally neutral states of Europe, or a national of one of the American republics which are all of them free from any connection with or participation in European political affairs.

Both Poland and Germany being sovereign governments, it is understood, of course, that upon resort to any one of the alternatives I suggest, each nation will agree to accord complete respect to the independence and territorial integrity of the other.

The people of the United States are as one in their opposition to policies of military conquest and domination. They are as one in rejecting the thesis that any ruler, or any people, possess the right to achieve their ends or objectives through the taking of action which will plunge countless millions of people into war and which will bring distress and suffering to every nation of the world, belligerent and neutral, when such ends and objectives, so far as they are just and reasonable, can be satisfied through processes of peaceful negotiation or by resort to judicial arbitration.

I appeal to you in the name of the people of the United States, and I believe in the name of peace-loving men and women everywhere, to agree to the solution of the controversies existing between your government and that of Poland through the adoption of one of the alternative methods I have proposed. I need hardly reiterate that should the governments of Germany and of Poland be willing to solve their differences in the peaceful manner suggested, the government of the United States still stands prepared to contribute its share to the solution of the problems which are endangering world peace in the form set forth in my message of April 14.

If his words couldn't bring peace to Europe, perhaps they would bring a measure of it to his fellow Americans. A few days after Hitler's troops invaded Poland on September 1, 1939, in a Fireside Chat, on September 3, FDR made the following remarks.

TONIGHT MY SINGLE duty is to speak to the whole of America.

Until 4:30 this morning I had hoped against hope that some miracle would prevent a devastating war in Europe and bring to an end the invasion of Poland by Germany.

For four long years a succession of actual wars and constant crises have shaken the entire world and have threatened in each case to bring on the gigantic conflict which is today unhappily a fact.

It is right that I should recall to your minds the consistent and at times successful efforts of your government in these crises to throw the full weight of the United States into the cause of peace. In spite of spreading wars, I think that we have every right and every reason to maintain as a national policy the fundamental moralities, the teachings of religion, and the continuation of efforts to restore peace—for some day, though the time may be distant, we can be of even greater help to a crippled humanity. . . .

You must master at the outset a simple but unalterable fact in modern foreign relations between nations. When peace has been broken anywhere, the peace of all countries everywhere is in danger.

It is easy for you and for me to shrug our shoulders and to say that conflicts taking place thousands of miles from the continental United States, and, indeed, thousands of miles from the whole American Hemisphere, do not seriously affect the Americas and that all the United States has to do is to ignore them and go about its own business. Passionately though we may desire detachment, we are forced to realize that every word that comes through the air, every ship that sails the sea, every battle that is fought does affect the American future.

Let no man or woman thoughtlessly or falsely talk of America sending its armies to European fields. At this moment there is being prepared a proclamation of American neutrality. This would have been done even if there had been no neutrality statute on the books, for this proclamation is in accordance with international law and in accordance with American policy.

This will be followed by a proclamation required by the existing Neutrality Act. And I trust that in the days to come our neutrality can be made a true neutrality.

It is of the utmost importance that the people of this country, with the best information in the world, think things through. The most dangerous enemies of American peace are those who, without well-rounded information on the whole broad subject of the past, the present, and the future, undertake to speak with assumed authority, to talk in terms of glittering generalities, to give to the nation assurances or prophecies which are of little present or future value.

I myself cannot and do not prophesy the course of events abroad—and the reason is that, because I have of necessity such a complete picture of what is going on in every part of the world, I do not dare to do so. And the other reason is that I think it is honest for me to be honest with the people of the United States.

I cannot prophesy the immediate economic effect of this new war on our nation, but I do say that no American has the moral right to profiteer at the expense either of his fellow citizens or of the men, the women, and the children who are living and dying in the midst of war in Europe. . . .

This nation will remain a neutral nation, but I cannot ask that every American remain neutral in thought as well. Even a neutral has a right to take account of facts. Even a neutral cannot be asked to close his mind or his conscience.

I have said not once, but many times, that I have seen war and that I hate war. I say that again and again.

I hope the United States will keep out of this war. I believe that it will. And I give you assurance and reassurance that every effort of your government will be directed toward that end.

As long as it remains within my power to prevent, there will be no blackout of peace in the United States.

FDR said privately, "I regret that Congress passed the Neutrality Act. I regret equally that I signed it." And on September 21, 1939, he made his feelings known to Congress, when he urged them to repeal the arms embargo to enable French and British forces to purchase munitions from the US.

MR. SPEAKER, MEMBERS of the Senate and House of Representatives: I have asked the Congress to reassemble in extraordinary session in order that it may consider and act on the amendment of certain legislation, which, in my best judgment, so alters the historic foreign policy of the United States that it impairs the peaceful relations of the United States with foreign nations.

At the outset I proceed on the assumption that every member of the Senate and of the House of Representatives, and every member of the executive branch of the government, including the president and his associates, personally and officially, are equally and without reservation in favor of such measures as will protect the neutrality, the safety, and the integrity of our country and at the same time keep us out of war.

Because I am wholly willing to ascribe an honorable desire for peace to those who hold different views from my own as to what those measures should be, I trust that these gentlemen will be sufficiently generous to ascribe equally lofty purposes to those with whom they disagree. Let no man or group in any walk of life assume exclusive protectorate over the future wellbeing of America, because I conceive that regardless of party or section the mantle of peace and of patriotism is wide enough to cover us all. Let no group assume the exclusive label of the "peace bloc." We all belong to it.

I have at all times kept the Congress and the American people informed of events and trends in foreign affairs. I now review them in a spirit of understatement.

Since 1931 the use of force instead of the council table has constantly increased in disputes between nations except in the Western Hemisphere, where in all those years there has been only one war, now happily terminated.

During those years also the building up of vast armies and navies and storehouses of war has proceeded abroad with growing speed and intensity. But, during these years, and extending back even to the days of the Kellogg-Briand Pact, the United States, has constantly, consistently, and conscientiously done all in its power to encourage peaceful settlements, to bring about reduction of armaments, and to avert threatened wars. We have done this not only because any war anywhere necessarily hurts American security and American prosperity, but because of the more important fact that any war anywhere retards the progress of morality and religion, and impairs the security of civilization itself.

For many years the primary purpose of our foreign policy has been that this nation and this government should strive to the utmost to aid in avoiding war among nations. But if and when war unhappily comes, the government and the nation must exert every possible effort to avoid being drawn into the war.

The executive branch of the government did its utmost, within our traditional policy of noninvolvement, to aid in averting the present appalling war. Having thus striven and failed, this government must lose no time or effort to keep our nation from being drawn into the war.

In my candid judgment we shall succeed in those efforts.

We are proud of the historical record of the United States and of all the Americans during all these years, because we have thrown every ounce of our influence for peace into the scale of peace.

Last January I told the Congress that "a war which threatened to envelop the world in flames has been averted, but it has become increasingly clear that peace is not assured."

By April new tensions had developed; a new crisis was in the making. Several nations with whom we had had friendly diplomatic and commercial relations had lost, or were in the process of losing, their independent identity and their very sovereignty.

During the spring and summer the trend was definitely toward further acts of military conquest and away from peace. As late as the end of July I spoke to members of the Congress about the definite possibility of war. I should have called it the probability of war.

Last January, also, I spoke to this Congress of the need for further warning of new threats of conquest, military and economic; of challenge to religion, to democracy, and to international good faith.

It was because of what I foresaw last January from watching the trend of foreign affairs and their probable effect upon us that I recommended to the Congress in July of this year that changes be enacted in our neutrality law.

The essentials for American peace in this war-torn world have not changed since last January or since last July. That is why I ask you again to reexamine our own legislation.

Beginning with the foundation of our constitutional government in the year 1789, the American policy in respect to belligerent nations, with one notable exception, has been based on international law. Be it remembered that what we call international law has always had as its primary objectives the avoidance of causes of war and the prevention of the extension of war.

The single exception to which I refer was the policy adopted by this nation during the Napoleonic Wars,

when, seeking to avoid involvement, we acted for some years under the so-called Embargo and Non-Intercourse Acts. That policy turned out to be a disastrous failure—first, because it brought our own nation close to ruin, and, secondly, because it was the major cause of bringing us into active participation in European wars in our own War of 1812. It is merely reciting history to recall to you that one of the results of the policy of embargo and nonintercourse was the burning in 1814 of part of this Capitol in which we are assembled today.

Our next deviation by statute from the sound principles of neutrality and peace through international law did not come for 130 years. It was the so-called Neutrality Act of 1935—only four years ago—an act continued in force by the Joint Resolution of May 1, 1937, despite grave doubts expressed as to its wisdom by many senators and representatives and by officials charged with the conduct of our foreign relations, including myself.

I regret that the Congress passed that act. I regret equally that I signed that act.

On July 14 of this year, I asked the Congress in the cause of peace and in the interest of real American neutrality and security, to take action to change that act.

I now ask again that such action be taken in respect to that part of the act which is wholly inconsistent with ancient precepts of the law of nations—the embargo provisions. I ask it because they are, in my opinion, most vitally dangerous to American neutrality, American security and, above all, American peace.

These embargo provisions, as they exist today, prevent the sale to a belligerent by an American factory of any completed implements of war, but they allow the sale of many types of uncompleted implements of war, as well as all kinds of general material and supplies. They, furthermore, allow such products of industry and agriculture to be taken in American flagships to belligerent nations. There in itself—under the present law—lies definite danger to our neutrality and our peace.

Let me set forth the present paradox of the existing legislation in its simplest terms. If, prior to 1935, a general war had broken out in Europe, the United States would have sold to, and bought from, belligerent nations such goods and products of all kinds as the belligerent nations, with their existing facilities and geographical situations, were able to buy from us or sell to us. This would have been the normal practice under the age-old doctrines of international law.

Our prior position accepted the facts of geography and of conditions of land power and sea power and air power alike, as they existed in all parts of the world.

If a war had broken out in Europe prior to 1935, there would have been no difference, for example, between our exports of sheets of aluminum and airplane wings; today there is an artificial legal difference.

Before 1935 there would have been no difference between the export of cotton and the export of gun cotton. Today there is.

Before 1935 there would have been no difference between the shipment of brass tubing in pipe form and brass tubing in shell form. Today there is.

Before 1935 there would have been no difference between the export of a motor truck and an armored motor truck. Today there is.

Let us be factual, let us recognize that a belligerent nation often needs wheat and lard and cotton for the survival of its population just as much as it needs antiaircraft guns and antisubmarine depth charges. Let those who seek to retain the present embargo position be wholly consistent. Let them seek new legislation to cut off cotton and cloth and copper and meat and wheat and a thousand other articles from all of the nations at war.

I seek a greater consistency through the repeal of the embargo provisions, and a return to international law. I seek reenactment of the historic and traditional American policy which, except for the disastrous interlude of the Embargo and Non-Intercourse Acts, has served us well from the very beginning of our constitutional existence.

It has been erroneously said that return to that policy might bring us nearer to war. I give to you my deep and unalterable conviction, based on years of experience as a worker in the field of international peace, that by the repeal of the embargo the United States will more probably remain at peace than if the law remains as it stands today. I say this because with the repeal of the embargo, this government clearly and definitely will insist that American citizens and American ships keep away from the immediate perils of the actual zones of conflict. . . .

When and if—I do not like even to mention the word "if," I would rather say "when"—repeal of the embargo is accomplished, certain other phases of policy reinforcing American safety should be considered. While nearly all of us are in agreement on their objectives, the only questions relate to method.

I believe that American merchant vessels should, as far as possible, be restricted from entering war zones. But, war zones may change so swiftly and so frequently in the days to come, that it is impossible to fix them permanently by act of Congress; specific legislation may prevent adjustment to constant and quick change. It seems, therefore, more practical to delimit the actual geography of the war zones through action of the State Department and administrative agencies.

The second objective is to prevent American citizens from traveling on belligerent vessels, or in danger areas. This can also be accomplished either by legislation, through continuance in force of certain provisions of existing law, or by proclamation making it clear to all Americans that any such travel is at their own risk.

The third objective, requiring the foreign buyer to take transfer of title in this country to commodities purchased by belligerents, is also a result that can be attained by legislation or substantially achieved through due notice by proclamation.

The fourth objective is the preventing of war credits to belligerents. This can be accomplished by maintaining in force existing provisions of law, or by proclamation making it clear that if credits are granted by American citizens to belligerents, our government will take no steps in the future to relieve them of risk or loss.

The result of these last two objectives will be to require all purchases to be made in cash, and all cargoes to be carried in the purchasers' own ships, at the purchasers' own risk.

Two other objectives have been amply attained by existing law, namely, regulating collection of funds in this country for belligerents, and the maintenance of a license system covering import and export of arms, ammunition, and implements of war. Under present enactments, such arms cannot be carried to belligerent countries on American vessels, and this provision should not be disturbed.

To those who say that this program would involve a step toward war on our part, I reply that it offers far greater safeguards than we now possess or have ever possessed, to protect American lives and property from danger. It is a positive program for giving safety. This means less likelihood of incidents and controversies which tend to draw us into conflict, as they unhappily did in the last World War. There lies the road to peace!

In respect to our own defense, you are aware that I have issued a proclamation setting forth "A National Emergency in Connection with Observance, Safeguarding, and Enforcement of Neutrality and the Strengthening of the National Defense within the Limits of Peacetime Authorization." This was done solely to make wholly constitutional and legal certain obviously necessary measures. I have authorized increases in the personnel of the army, the navy, the Marine Corps, and the Coast Guard, which will bring all four of them to a total still below peacetime strength as authorized by the Congress.

I have authorized the State Department to use, for the repatriation of Americans caught in the war zone, the sum of five hundred thousand dollars already authorized by the Congress.

I have authorized the addition of 150 persons to the Department of Justice to be used in the protection of the United States against subversive foreign activities within our borders.

At this time I ask for no further authority from the Congress. At this time I see no need for further executive action under the proclamation of limited national emergency.

Therefore, I see no impelling reason for the consideration of other legislation at this extraordinary session of the Congress.

It is, of course, possible that in the months to come

unforeseen needs for further legislation may develop but they are not imperative today. . . .

I should like to be able to offer the hope that the shadow over the world might swiftly pass. I cannot. The facts compel my stating, with candor, that darker periods may lie ahead. The disaster is not of our making; no act of ours engendered the forces which assault the foundations of civilization. Yet we find ourselves affected to the core; our currents of commerce are changing, our minds are filled with new problems, our position in world affairs has already been altered.

In such circumstances our policy must be to appreciate in the deepest sense the true American interest. Rightly considered, this interest is not selfish. Destiny first made us, with our sister nations on this hemisphere, joint heirs of European culture. Fate seems now to compel us to assume the task of helping to maintain in the Western world a citadel wherein that civilization may be kept alive. The peace, the integrity, and the safety of the Americas—these must be kept firm and serene.

In a period when it is sometimes said that free discussion is no longer compatible with national safety, may you by your deeds show the world that we of the United States are one people, of one mind, one spirit, one clear resolution, walking before God in the light of the living.

To get around the Neutrality Act, Roosevelt decided that if war were not formally declared on Germany by the Western powers, he shouldn't be forced to invoke the arms embargo required by the Act. "Let them call it defensive measures or anything plausible, but avoid actual declaration of war."

Roosevelt revealed just how far he was willing to go to help the French and British forces should they be "forced to war." Battle should be waged "purely by blockade and in a defensive manner. . . . The Powers should close their own frontiers to Germany, stand on an armed defensive, and call on all other states adjoining Germany to adopt the same line of nonintercourse. The blockade line should be drawn down the middle of the North Sea, through the Channel to Gibraltar, and the Mediterranean should be closed at the Suez Canal. . . . Blockade must be based on [the] loftiest humanitarian grounds, and on the desire to wage hostilities with [a] minimum of suffering and the least possible loss of life and property, and yet to bring [the] enemy to his knees." Roosevelt was certain that the American people would approve of a blockade "if its humanitarian purpose were strongly emphasized."

In the summer of 1940, the president would still be seeking ways around the act. Here is Roosevelt's own record of a cabinet meeting on August 2.

AT CABINET MEETING, in afternoon, long discussion in regard to devising ways and means to sell directly or indirectly fifty or sixty World War old destroyers to Great Britain. It was the general opinion, without any dissenting voice, that the survival of the British Isles under German attack might very possibly depend on their getting these destroyers.

It was agreed that legislation to accomplish this was necessary.

It was agreed that such legislation if asked for by me without any preliminaries would meet with defeat or interminable delay in reaching a vote.

It was agreed that the British be approached through Lord Lothian to find out if they would agree to give positive assurances that the British Navy, in the event of German success in Great Britain, would not under any conceivable circumstances fall into the hands of the Germans and that if such assurances could be received and made public, the opposition in the Congress would be greatly lessened. I suggested that we try to get further assurance from the British that the ships of their Navy would not be sunk but would sail for North America or British Empire ports where they would remain afloat and available.

I stressed the point that in all probability the legislation would fail if it had substantially unanimous Republican opposition—and that the crux of the matter lay in the vote of the Republican minority in each House. I stressed the importance of having the issue acted on without regard to party politics in any way.

Ultimately, Roosevelt would find the support he needed. The arms embargo provision of the Neutrality Act would be overturned. And in 1941, Congress would pass the Lend-Lease program, giving FDR the power to supply arms to any nation whose defense was perceived to be intrinsically tied to America's security.

Roosevelt was very slow in deciding whether to run for an unprecedented third term. "I do not want to run," he told Treasury Secretary Henry Morgenthau, Jr., "unless . . . things get very, very much worse in Europe." But in an extremely dramatic radio address to the Democratic National Convention on July 19, 1940, Roosevelt ended the suspense regarding his intentions.

MEMBERS OF THE Convention—My Friends: It is very late; but I have felt that you would rather that I speak to you now than wait until tomorrow.

It is with a very full heart that I speak tonight. I must confess that I do so with mixed feelings—because I find myself, as almost everyone does sooner or later in his lifetime, in a conflict between deep personal desire for retirement on the one hand, and that quiet, invisible thing called "conscience" on the other.

Because there are self-appointed commentators and interpreters who will seek to misinterpret or question motives, I speak in a somewhat personal vein; and I must trust to the good faith and common sense of the American people to accept my own good faith—and to do their own interpreting.

When, in 1936, I was chosen by the voters for a second time as president, it was my firm intention to turn over the responsibilities of government to other hands at the end of my term. That conviction remained with me. Eight years in the presidency, following a period of bleak depression, and covering one world crisis after another, would normally entitle any man to the relaxation that comes from honorable retirement.

During the spring of 1939, world events made it clear to all but the blind or the partisan that a great war in Europe had become not merely a possibility but a probability, and that such a war would of necessity deeply affect the future of this nation.

When the conflict first broke out last September, it was still my intention to announce clearly and simply, at an early date, that under no conditions would I accept reelection. This fact was well known to my friends, and I think was understood by many citizens.

It soon became evident, however, that such a public statement on my part would be unwise from the point of view of sheer public duty. As president of the United States, it was my clear duty, with the aid of the Congress, to preserve our neutrality, to shape our program of defense, to meet rapid changes, to keep our domestic affairs adjusted to shifting world conditions, and to sustain the policy of the Good Neighbor.

It was also my obvious duty to maintain to the utmost the influence of this mighty nation in our effort to prevent the spread of war, and to sustain by all legal means those governments threatened by other governments which had rejected the principles of democracy.

Swiftly moving foreign events made necessary swift action at home and beyond the seas. Plans for national defense had to be expanded and adjusted to meet new forms of warfare. American citizens and their welfare had to be safeguarded in many foreign zones of danger. National unity in the United States became a crying essential in the face of the development of unbelievable types of espionage and international treachery.

Every day that passed called for the postponement of personal plans and partisan debate until the latest possible moment. The normal conditions under which I would have made public declaration of my personal desires were wholly gone.

And so, thinking solely of the national good and of the international scene, I came to the reluctant conclusion that such declaration should not be made before the national convention. It was accordingly made to you within an hour after the permanent organization of this convention.

Like any other man, I am compelled by the honor you have done me. But I know you will understand the spirit in which I say that no call of party alone would prevail upon me to accept reelection to the presidency.

The real decision to be made in these circumstances is not the acceptance of a nomination, but rather an ultimate willingness to serve if chosen by the electorate of the United States. . . .

Lying awake, as I have, on many nights, I have asked myself whether I have the right as commander in chief of the army and navy, to call on men and women to serve their country or to train themselves to serve and, at the same time, decline to serve my country in my own personal capacity, if I am called upon to do so by the people of my country.

In times like these—in times of great tension, of great crisis—the compass of the world narrows to a single fact. The fact which dominates our world is the fact of armed aggression, the fact of successful armed aggression, aimed at the form of government, the kind of society that we in the United States have chosen and established for ourselves. It is a fact which no one longer doubts, which no one is longer able to ignore.

It is not an ordinary war. It is a revolution imposed by force of arms, which threatens all men everywhere. It is a revolution which proposes not to set men free but to reduce them to slavery—to reduce them to slavery in the interest of a dictatorship which has already shown the nature and the extent of the advantage which it hopes to obtain.

That is the fact which dominates our world and which

dominates the lives of all of us, each and every one of us. In the face of the danger which confronts our time, no individual retains, or can hope to retain, the right of personal choice which free men enjoy in times of peace. He has a first obligation to serve in the defense of our institutions of freedom—a first obligation to serve his country in whatever capacity his country finds him useful.

Like most men of my age, I had made plans for myself, plans for a private life of my own choice and for my own satisfaction, a life of that kind to begin in January 1941. These plans, like so many other plans, had been made in a world which now seems as distant as another planet. Today all private plans, all private lives, have been in a sense repealed by an overriding public danger. In the face of that public danger all those who can be of service to the Republic have no choice but to offer themselves for service in those capacities for which they may be fitted.

Those, my friends, are the reasons why I have had to admit to myself, and now to state to you, that my conscience will not let me turn my back upon a call to service.

The right to make that call rests with the people through the American method of a free election. Only the people themselves can draft a president. If such a draft should be made upon me, I say to you, in the utmost simplicity, I will, with God's help, continue to serve with the best of my ability and with the fullness of my strength.

In some respects, as I think my good wife suggested an hour or so ago, the next few months will be different from the usual national campaigns of recent years.

Most of you know how important it is that the president of the United States in these days remain close to the seat of government. Since last summer, I have been compelled to abandon proposed journeys to inspect many of our great national projects from the Alleghenies to the Pacific Coast.

Events move so fast in other parts of the world that it has become my duty to remain either in the White House itself or at some nearby point where I can reach Washington and even Europe and Asia by direct telephone—where, if need be, I can be back at my desk in the space of a very few hours. And in addition, the splendid work of the new defense machinery will require me to spend vastly more time in conference with the responsible administration heads under me. Finally, the added task which the present crisis has imposed also upon the Congress, compelling them to forego their usual adjournment, calls for constant cooperation between the executive and legislative branches, to the efficiency of which I am glad indeed now to pay tribute.

I do expect, of course, during the coming months to make my usual periodic reports to the country through the medium of press conferences and radio talks. I shall not have the time or the inclination to engage in purely political debate. But I shall never be loath to call the attention of the nation to deliberate or unwitting falsifications of fact, which are sometimes made by political candidates.

I have spoken to you in a very informal and personal way. The exigencies of the day require, however, that I also talk with you about things which transcend any personality and go very deeply to the roots of American civilization.

Our lives have been based on those fundamental freedoms and liberties which we Americans have cherished for a century and a half. The establishment of them and the preservation of them in each succeeding generation have been accomplished through the processes of free elective government—the democratic-republican form, based on the representative system and the coordination of the executive, the legislative, and the judicial branches.

The task of safeguarding our institutions seems to me to be twofold. One must be accomplished, if it becomes necessary, by the armed defense forces of the nation. The other, by the united efforts of the men and women of the country to make our federal and state and local governments responsive to the growing requirements of modern democracy.

There have been occasions, as we remember, when reactions in the march of democracy have set in, and the forward-looking progress has seemed to stop.

But such periods have been followed by liberal and progressive times which have enabled the nation to catch up with new developments in fulfilling new human needs. Such a time has been the past seven years. Because we had seemed to lag in previous years, we have had to develop, speedily and efficiently, the answers to aspirations which had come from every state and every family in the land.

We have sometimes called it social legislation; we have sometimes called it legislation to end the abuses of the past; we have sometimes called it legislation for human security; and we have sometimes called it legislation to better the condition of life of the many millions of our fellow citizens who could not have the essentials of life or hope for an American standard of living.

Some of us have labeled it a wider and more equitable distribution of wealth in our land. It has included among its aims to liberalize and broaden the control of vast industries—lodged today in the hands of a relatively small group of individuals of very great financial power.

But all of these definitions and labels are essentially the expression of one consistent thought. They represent a constantly growing sense of human decency, human decency throughout our nation.

This sense of human decency is happily confined to no group or class. You find it in the humblest home. You find it among those who toil, and among the shopkeepers and the farmers of the nation. You find it, to a growing degree, even among those who are listed in that top group which has so much control over the industrial and financial structure of the nation. Therefore, this urge of humanity can by no means be labeled a war of class against class. It is rather a war in which all classes are joining in the interest of a sound and enduring democracy.

I do not believe for a moment, and I know that you do not believe either, that we have fully answered all the needs of human security. But we have covered much of that road.

But we all know that our progress at home and in the other American nations toward this realization of a better human decency—progress along free lines—is gravely endangered by what is happening on other continents. In Europe, many nations, through dictatorships or invasions, have been compelled to abandon normal democratic processes. They have been compelled to adopt forms of government which some call "new and efficient."

They are not new, my friends, they are only a relapse—a relapse into ancient history. The omnipotent rulers of the greater part of modern Europe have guaranteed efficiency, and work, and a type of security.

But the slaves who built the pyramids for the glory of the dictator pharaohs of Egypt had that kind of security, that kind of efficiency, that kind of corporative state.

So did the inhabitants of that world which extended from Britain to Persia under the undisputed rule of the proconsuls sent out from Rome.

So did the henchmen, the tradesmen, the mercenaries, and the slaves of the feudal system which dominated Europe a thousand years ago.

So did the people of those nations of Europe who received their kings and their governments at the whim of the conquering Napoleon.

Whatever its new trappings and new slogans, tyranny is the oldest and most discredited rule known to history. And whenever tyranny has replaced a more human form of government, it has been due more to internal causes than external. Democracy can thrive only when it enlists the devotion of those whom Lincoln called the common people. Democracy can hold that devotion only when it adequately respects their dignity by so ordering society as to assure to the masses of men and women reasonable security and hope for themselves and for their children.

The government of the United States for the past seven years has had the courage openly to oppose by every peaceful means the spread of the dictator form of government. If our government should pass to other hands next January—untried hands, inexperienced hands—we can merely hope and pray that they will not substitute appeasement and compromise with those who seek to destroy all democracies everywhere, including here.

I would not undo, if I could, the efforts I made to prevent war from the moment it was threatened and to restrict the area of carnage, down to the last minute. I do not now soften the condemnation expressed by Secretary Hull and myself from time to time for the acts of aggression that have wiped out ancient liberty-loving, peace-pursuing countries which had scrupulously maintained neutrality. I do not recant the sentiments of sympathy with all free peoples resisting such aggression, or begrudge the material aid that we have given to them. I do not regret my consistent endeavor to awaken this country to the menace for us and for all we hold dear.

I have pursued these efforts in the face of appeaser fifth columnists who charged me with hysteria and warmongering. But I felt it my duty, my simple, plain, inescapable duty to arouse my countrymen to the danger of the new forces let loose in the world.

So long as I am president, I will do all I can to insure that that foreign policy remain our foreign policy.

All that I have done to maintain the peace of this country and to prepare it morally, as well as physically, for whatever contingencies may be in store, I submit to the judgment of my countrymen.

We face one of the great choices of history.

It is not alone a choice of government by the people versus dictatorship.

It is not alone a choice of freedom versus slavery.

It is not alone a choice between moving forward or falling back.

It is all of these rolled into one.

It is the continuance of civilization as we know it versus the ultimate destruction of all that we have held dear—religion against godlessness; the ideal of justice against the practice of force; moral decency versus the firing squad; courage to speak out, and to act, versus the false lullaby of appeasement.

But it has been well said that a selfish and greedy people cannot be free.

The American people must decide whether these

things are worth making sacrifices of money, of energy, and of self. They will not decide by listening to mere words or by reading mere pledge, interpretations, and claims. They will decide on the record—the record as it has been made, the record of things as they are.

The American people will sustain the progress of a representative democracy, asking the Divine Blessing as they face the future with courage and with faith.

As important to America as FDR's third—and fourth—terms would prove to be, nonethless, Congress passed the Twenty-second Amendment in 1951, limiting all future presidents to two terms or ten years in office total.

On September 16, 1940, Roosevelt took a firm step himself toward "sustain[ing] the progress of representative democracy" when he made the following statement on the adoption of the peacetime Selective Service:

❧ AMERICA HAS ADOPTED Selective Service in time of peace, and, in doing so, has broadened and enriched our basic concept of citizenship. Beside the clear democratic ideals of equal rights, equal privileges, and equal opportunities, we have set forth the underlying other duties, obligations, and responsibilities of equal service.

In thus providing for national defense, we have not carved a new and uncharted trail in the history of our democratic institutions. On the contrary, we have merely reasserted an old and accepted principle of democratic government. The militia system, the self-armed citizenry with the obligation of military service incumbent upon every free man, has its roots in the old common law. It was brought to this continent by our forefathers. It was an accepted institution in colonial days. At the time of the adoption of the federal Constitution, nine of the thirteen states explicitly provided for universal service in their basic laws.

In those days, little was required in the way of equipment and training for the man in arms. The average American had his flintlock and knew how to use it. In addition, he was healthy, strong, and accustomed to hardship. When he reported for military duty, he brought with him his musket and his powder horn. His daily life inured him to the rigors of warfare.

Today, the art of war calls for a wide variety of technical weapons. Modern life does not emphasize the qualities demanded of soldiers. Moreover, behind the armed forces, we must have a munitions industry as a part of an economic system capable of providing the fighting man with his full requirements of arms and equipment. Many individuals, therefore, may serve their country best by holding their posts on the production line. The object of Selective Service is to provide men for our army and navy and at the same time disturb as little as possible the normal life of the union.

Selective Service consists of four steps, which singly and in the group have been developed to operate with the fairness and justice characteristic of free, democratic institutions. These steps are: registration, classification, selection, and induction.

Wednesday, October 16, has been set aside, on which day every male between twenty-one and thirty-five, inclusive, will be expected to report to a neighborhood precinct to fill out a registration card and a registration certificate.

The certificate issued to the individual will be carried by him as a testimonial to his acceptance of the fundamental obligation of citizenship. The registration card will be forwarded to the county clerk or similar official and will be delivered by him to the local Selective Service board. These boards, consisting of three men, each appointed by the president, upon recommendations of the state governors, will be set up in more than 6,000 communities. When the states notify the national director of Selective Service that all the local boards have completed this work, a national drawing by lot will determined the order of priority of the registrants in each local board area. The national priority list will be furnished to the local boards and the corresponding order of selection will be entered on the registration cards in their custody.

The priority established by the drawing will determine the order in which the questionnaires will be mailed to the registrants. Upon receipt of these questionnaires, the registrants will enter on these forms pertinent facts on the basis of which their final classification will be determined.

In the military service, Americans from all walks of life—rich and poor, country-bred and city-raised, farmer, student, manual laborer, and white-collar worker—will learn to live side by side, to depend upon each other in military drills and maneuvers, and to appreciate each other's dignity as American citizens.

Universal service will bring not only greater preparedness to meet the threat of war, but a wider distribution of tolerance and understanding to enjoy the blessings of peace.

Naturally, such measures led to FDR being charged that he would lead his country into war. He responded to those of

Wendell Willkie—his toughest opponent to date—with vehemence: "A deliberate falsification of fact. . . . [Willkie and the Republicans are] playing politics with national defense." And on the campaign trail, Roosevelt hammered home his position. On Columbus Day he addressed the nation: "No combination of dictator countries of Europe and Asia will stop the help we are giving to almost the last free people now fighting to hold them at bay. . . . We will continue to pile up our defenses and our armaments. We will continue to help those who resist aggression, and who now hold the oppressors far from our shores. . . . The people of the United States, the people of all the Americas, reject the doctrine of appeasement. They recognize it for what it is—a major weapon of the aggressor nations."

Although the number of isolationists in America was sizable, and their opposition to any *form of involvement—even in the form of minimal support for nations victimized by Hitler and his allies—was organized and pronounced, FDR ever stayed the course.*

Close to the November 1940 election, to avert a seaborne German invasion, Winston Churchill all but begged Roosevelt to provide American destroyers. Roosevelt acted quickly. "Congress is going to raise hell about this," he told his secretary Grace Tully, "but even another day's delay may mean the end of civilization."

Roosevelt made his eighth annual message to Congress on January 6, 1941. In it, he outlined a way to help the increasingly strained Allies finance the war. He also explicated the famous "Four Freedoms."

❧ MR. SPEAKER, MEMBERS of the Seventy-seventh Congress: I address you, the members of the Seventy-seventh

Congress, at a moment unprecedented in the history of the Union. I use the word *unprecedented* because at no previous time has American security been as seriously threatened from without as it is today.

Since the permanent formation of our government under the Constitution, in 1789, most of the periods of crisis in our history have related to our domestic affairs. Fortunately, only one of these—the four-year War Between the States—ever threatened our national unity. Today, thank God, 130 million Americans, in forty-eight states, have forgotten points of the compass in our national unity.

It is true that prior to 1914 the United States often had been disturbed by events in other continents. We had even engaged in two wars with European nations and in a number of undeclared wars in the West Indies, in the Mediterranean, and in the Pacific for the maintenance of American rights and for the principles of peaceful commerce. But in no case had a serious threat been raised against our national safety or our continued independence.

What I seek to convey is the historic truth that the United States as a nation has at all times maintained clear, definite opposition to any attempt to lock us in behind an ancient Chinese wall while the procession of civilization went past. Today, thinking of our children and of their children, we oppose enforced isolation for ourselves or for any other part of the Americas.

That determination of ours, extending over all these years, was proved, for example, during the quarter century of wars following the French Revolution.

While the Napoleonic struggles did threaten interests of the United States because of the French foothold in the

West Indies and in Louisiana, and while we engaged in the War of 1812 to vindicate our right to peaceful trade, it is nevertheless clear that neither France nor Great Britain, nor any other nation, was aiming at domination of the whole world.

In like fashion, from 1815 to 1914—ninety-nine years—no single war in Europe or in Asia constituted a real threat against our future or against the future of any other American nation.

Except in the Maximilian interlude in Mexico, no foreign power sought to establish itself in this hemisphere; and the strength of the British fleet in the Atlantic has been a friendly strength. It is still a friendly strength.

Even when the World War broke out in 1914, it seemed to contain only small threat of danger to our own American future. But, as time went on, the American people began to visualize what the downfall of democratic nations might mean to our own democracy.

We need not overemphasize imperfections in the Peace of Versailles. We need not harp on failure of the democracies to deal with problems of world reconstruction. We should remember that the Peace of 1919 was far less unjust than the kind of "pacification" which began even before Munich, and which is being carried on under the new order of tyranny that seeks to spread over every continent today. The American people have unalterably set their faces against that tyranny.

Every realist knows that the democratic way of life is at this moment being directly assailed in every part of the world—assailed either by arms, or by secret spreading of poisonous propaganda by those who seek to destroy unity and promote discord in nations that are still at peace.

During sixteen long months this assault has blotted

out the whole pattern of democratic life in an appalling number of independent nations, great and small. The assailants are still on the march, threatening other nations, great and small.

Therefore, as your president, performing my constitutional duty to "give to the Congress information of the state of the Union," I find it, unhappily, necessary to report that the future and the safety of our country and of our democracy are overwhelmingly involved in events far beyond our borders.

Armed defense of democratic existence is now being gallantly waged in four continents. If that defense fails, all the population and all the resources of Europe, Asia, Africa, and Australia will be dominated by the conquerors. Let us remember that the total of those populations and their resources in those four continents greatly exceeds the sum total of the population and the resources of the whole of the Western Hemisphere—many times over.

In times like these it is immature—and incidentally, untrue—for anybody to brag that an unprepared America, single-handedly, and with one hand tied behind its back, can hold off the whole world.

No realistic American can expect from a dictator's peace international generosity, or return of true independence, or world disarmament, or freedom of expression, or freedom of religion—or even good business. . . .

I have recently pointed out how quickly the tempo of modern warfare could bring into our very midst the physical attack which we must eventually expect if the dictator nations win this war.

There is much loose talk of our immunity from immediate and direct invasion from across the seas. Obviously, as long as the British navy retains its power, no such danger exists. Even if there were no British navy, it is not probable that any enemy would be stupid enough to attack us by landing troops in the United States from across thousands of miles of ocean, until it had acquired strategic bases from which to operate.

But we learn much from the lessons of the past years in Europe—particularly the lesson of Norway, whose essential seaports were captured by treachery and surprise built up over a series of years.

The first phase of the invasion of this hemisphere would not be the landing of regular troops. The necessary strategic points would be occupied by secret agents and their dupes—and great numbers of them are already here, and in Latin America.

As long as the aggressor nations maintain the offensive, they—not we—will choose the time and the place and the method of their attack.

That is why the future of all the American republics is today in serious danger. . . .

Our national policy is this:

First, by an impressive expression of the public will and without regard to partisanship, we are committed to all-inclusive national defense.

Second, by an impressive expression of the public will and without regard to partisanship, we are committed to full support of all those resolute peoples, everywhere, who are resisting aggression and are thereby keeping war away from our hemisphere. By this support, we express our determination that the democratic cause shall prevail; and we strengthen the defense and the security of our own nation.

Third, by an impressive expression of the public will and without regard to partisanship, we are committed to the proposition that principles of morality and considerations for our own security will never permit us to acquiesce in a peace dictated by aggressors and sponsored by appeasers. We know that enduring peace cannot be bought at the cost of other people's freedom. . . .

Therefore, the immediate need is a swift and driving increase in our armament production.

Leaders of industry and labor have responded to our summons. Goals of speed have been set. In some cases these goals are being reached ahead of time; in some cases we are on schedule; in other cases there are slight but not serious delays; and in some cases—and I am sorry to say very important cases—we are all concerned by the slowness of the accomplishment of our plans.

The army and navy, however, have made substantial progress during the past year. Actual experience is improving and speeding up our methods of production with every passing day.

And today's best is not good enough for tomorrow.

I am not satisfied with the progress thus far made. The men in charge of the program represent the best in training, in ability, and in patriotism. They are not satisfied with the progress thus far made. None of us will be satisfied until the job is done. . . .

To change a whole nation from a basis of peacetime production of implements of peace to a basis of wartime production of implements of war is no small task. And the greatest difficulty comes at the beginning of the program, when new tools, new plant facilities, new assembly lines, and new shipways must first be constructed before the actual materiel begins to flow steadily and speedily from them. . . .

New circumstances are constantly begetting new

needs for our safety. I shall ask this Congress for greatly increased new appropriations and authorizations to carry on what we have begun.

I also ask this Congress for authority and for funds sufficient to manufacture additional munitions and war supplies of many kinds, to be turned over to those nations which are now in actual war with aggressor nations.

Our most useful and immediate role is to act as an arsenal for them as well as for ourselves. They do not need manpower, but they do need billions of dollars worth of the weapons of defense.

The time is near when they will not be able to pay for them all in ready cash. We cannot, and we will not, tell them that they must surrender, merely because of present inability to pay for the weapons which we know they must have.

I do not recommend that we make them a loan of dollars with which to pay for these weapons—a loan to be repaid in dollars.

I recommend that we make it possible for those nations to continue to obtain war materials in the United States, fitting their orders into our own program. Nearly all their materiel would, if the time ever came, be useful for our own defense. . . .

For what we send abroad, we shall be repaid within a reasonable time following the close of hostilities, in similar materials, or, at our option, in other goods of many kinds, which they can produce and which we need.

Let us say to the democracies: "We Americans are vitally concerned in your defense of freedom. We are putting forth our energies, our resources, and our organizing powers to give you the strength to regain and maintain a free world. We shall send you, in ever-increasing numbers, ships, planes, tanks, guns. This is our purpose and our pledge."

In fulfillment of this purpose we will not be intimidated by the threats of dictators that they will regard as a breach of international law or as an act of war our aid to the democracies which dare to resist their aggression. Such aid is not an act of war, even if a dictator should unilaterally proclaim it so to be.

When the dictators—if the dictators—are ready to make war upon us, they will not wait for an act of war on our part. They did not wait for Norway or Belgium or the Netherlands to commit an act of war. . . .

The nation takes great satisfaction and much strength from the things which have been done to make its people conscious of their individual stake in the preservation of democratic life in America. Those things have toughened the fiber of our people, have renewed their faith and strengthened their devotion to the institutions we make ready to protect.

Certainly this is no time for any of us to stop thinking about the social and economic problems which are the root cause of the social revolution which is today a supreme factor in the world.

For there is nothing mysterious about the foundations of a healthy and strong democracy. The basic things expected by our people of their political and economic systems are simple. They are:

Equality of opportunity for youth and for others.

Jobs for those who can work.

Security for those who need it.

The ending of special privilege for the few.

The preservation of civil liberties for all.

The enjoyment of the fruits of scientific progress in a wider and constantly rising standard of living.

These are the simple, basic things that must never be lost sight of in the turmoil and unbelievable complexity of our modern world. The inner and abiding strength of our economic and political systems is dependent upon the degree to which they fulfill these expectations.

Many subjects connected with our social economy call for immediate improvement.

As examples:

We should bring more citizens under the coverage of old-age pensions and unemployment insurance.

We should widen the opportunities for adequate medical care.

We should plan a better system by which persons deserving or needing gainful employment may obtain it.

I have called for personal sacrifice. I am assured of the willingness of almost all Americans to respond to that call. . . .

In the future days, which we seek to make secure, we look forward to a world founded upon four essential human freedoms.

The first is freedom of speech and expression—everywhere in the world.

The second is freedom of every person to worship God in his own way—everywhere in the world.

The third is freedom from want—which, translated into world terms, means economic understandings which will secure to every nation a healthy peacetime life for its inhabitants—everywhere in the world.

The fourth is freedom from fear which, translated into world terms, means a worldwide reduction of armaments to such a point and in such a thorough fashion that no nation will be in a position to commit an act of physical aggression against any neighbor—anywhere in the world.

That is no vision of a distant millennium. It is a definite basis for a kind of world attainable in our own time and generation. That kind of world is the very antithesis of the so-called new order of tyranny which the dictators seek to create with the crash of a bomb.

To that new order we oppose the greater conception—the moral order. A good society is able to face schemes of world domination and foreign revolutions alike without fear.

Since the beginning of our American history, we have been engaged in change—in a perpetual peaceful revolution, a revolution which goes on steadily, quietly adjusting itself to changing conditions—without the concentration camp or the quick-lime in the ditch. The world order which we seek is the cooperation of free countries, working together in a friendly, civilized society.

This nation has placed its destiny in the hands and heads and hearts of its millions of free men and women; and its faith in freedom under the guidance of God. Freedom means the supremacy of human rights everywhere. Our support goes to those who struggle to gain those rights or keep them. Our strength is our unity of purpose.

To that high concept there can be no end save victory.

A few weeks later, on January 20, 1941, he gave his stirring third Inaugural Address; which took a different tack by appealing directly to the American heart.

ON EACH NATIONAL day of inauguration since 1789, the people have renewed their sense of dedication to the United States.

In Washington's day the task of the people was to create and weld together a nation.

In Lincoln's day the task of the people was to preserve that nation from disruption from within.

In this day the task of the people is to save that nation and its institutions from disruption from without.

To us there has come a time, in the midst of swift happenings, to pause for a moment and take stock to recall what our place in history has been, and to rediscover what we are and what we may be. If we do not, we risk the real peril of isolation, the real peril of inaction.

Lives of nations are determined not by the count of years, but by the lifetime of the human spirit. The life of a man is threescore years and ten: a little more, a little less. The life of a nation is the fullness of the measure of its will to live.

There are men who doubt this. There are men who believe that democracy, as a form of government and a frame of life, is limited or measured by a kind of mystical and artificial fate—that, for some unexplained reason, tyranny and slavery have become the surging wave of the future and that freedom is an ebbing tide.

But we Americans know that this is not true.

Eight years ago, when the life of this Republic seemed frozen by a fatalistic terror, we proved that this is not true. We were in the midst of shock, but we acted. We acted quickly, boldly, decisively.

These later years have been living years—fruitful years for the people of this democracy. For they have brought to us greater security and, I hope, a better understanding that life's ideals are to be measured in other than material things.

Most vital to our present and to our future is this experience of a democracy which successfully survived crisis at home; put away many evil things; built new structures on enduring lines; and, through it all, maintained the fact of its democracy.

For action has been taken within the three-way framework of the Constitution of the United States. The coordinate branches of the government continue freely to function. The Bill of Rights remains inviolate. The freedom of elections is wholly maintained. Prophets of the downfall of American democracy have seen their dire predictions come to naught.

No, democracy is not dying.

We know it because we have seen it revive—and grow.

We know it cannot die—because it is built on the unhampered initiative of individual men and women joined together in a common enterprise, an enterprise undertaken and carried through by the free expression of a free majority.

We know it because democracy alone, of all forms of government, enlists the full force of men's enlightened will.

We know it because democracy alone has constructed an unlimited civilization capable of infinite progress in the improvement of human life.

We know it because, if we look below the surface, we sense it still spreading on every continent—for it is the most humane, the most advanced, and in the end the most unconquerable of all forms of human society.

A nation, like a person, has a body—a body that must

be fed and clothed and housed, invigorated and rested, in a manner that measures up to the standards of our time.

A nation, like a person, has a mind—a mind that must be kept informed and alert, that must know itself, that understands the hopes and the needs of its neighbors—all the other nations that live within the narrowing circle of the world.

A nation, like a person, has something deeper, something more permanent, something larger than the sum of all its parts. It is that something which matters most to its future, which calls forth the most sacred guarding of its present.

It is a thing for which we find it difficult—even impossible—to hit upon a single, simple word.

And yet, we all understand what it is: the spirit, the faith of America. It is the product of centuries. It was born in the multitudes of those who came from many lands, some of high degree, but mostly plain people who sought here, early and late, to find freedom more freely.

The democratic aspiration is no mere recent phase in human history. It is human history. It permeated the ancient life of early peoples. It blazed anew in the Middle Ages. It was written in Magna Carta.

In the Americas its impact has been irresistible. America has been the New World in all tongues, and to all peoples, not because this continent was a new-found land, but because all those who came here believed they could create upon this continent a new life—a life that should be new in freedom.

Its vitality was written into our own Mayflower Compact, into the Declaration of Independence, into the Constitution of the United States, into the Gettysburg Address.

Those who first came here to carry out the longings of their spirit, and the millions who followed, and the stock that sprang from them—all have moved forward constantly and consistently toward an ideal which in itself has gained stature and clarity with each generation.

The hopes of the Republic cannot forever tolerate either undeserved poverty or self-serving wealth.

We know that we still have far to go; that we must more greatly build the security and the opportunity and the knowledge of every citizen, in the measure justified by the resources and the capacity of the land.

But it is not enough to achieve these purposes alone. It is not enough to clothe and feed the body of this nation, to instruct and inform its mind. For there is also the spirit. And of the three, the greatest is the spirit.

Without the body and the mind, as all men know, the nation could not live.

But if the spirit of America were killed, even though the nation's body and mind, constricted in an alien world, lived on, the America we know would have perished.

That spirit—that faith—speaks to us in our daily lives in ways often unnoticed, because they seem so obvious. It speaks to us here in the capital of the nation. It speaks to us through the processes of governing in the sovereignties of forty-eight states. It speaks to us in our counties, in our cities, in our towns, and in our villages. It speaks to us from the other nations of the hemisphere, and from those across the seas—the enslaved, as well as the free. Sometimes we fail to hear or heed these voices of freedom because to us the privilege of our freedom is such an old, old story.

The destiny of America was proclaimed in words of prophecy spoken by our first president in his first inaugural in 1789—words almost directed, it would seem, to this year of 1941: "The preservation of the sacred fire of liberty and the destiny of the republican model of government are justly considered . . . deeply . . . finally, staked on the experiment intrusted to the hands of the American people."

If you and I in this later day lose that sacred fire—if we let it be smothered with doubt and fear—then we shall reject the destiny which Washington strove so valiantly and so triumphantly to establish. The preservation of the spirit and faith of the nation does, and will, furnish the highest justification for every sacrifice that we may make in the cause of national defense.

In the face of great perils never before encountered, our strong purpose is to protect and to perpetuate the integrity of democracy.

For this we muster the spirit of America, and the faith of America.

We do not retreat. We are not content to stand still. As Americans, we go forward, in the service of our country, by the will of God.

On May 27, 1941, Roosevelt made a particularly rousing radio address, one far less majestic than his inaugural address, but no less moving for its "cold' recitation of the facts.

. . . THE PRESSING PROBLEMS that confront us are military and naval problems. We cannot afford to approach them from the point of view of wishful thinkers or sentimentalists. What we face is cold, hard fact.

The first and fundamental fact is that what started as

a European war has developed, as the Nazis always intended it should develop, into a world war for world domination.

Adolf Hitler never considered the domination of Europe as an end in itself. European conquest was but a step toward ultimate goals in all the other continents. It is unmistakably apparent to all of us that, unless the advance of Hitlerism is forcibly checked now, the Western Hemisphere will be within range of the Nazi weapons of destruction.

For our own defense we have accordingly undertaken certain obviously necessary measures:

First, we have joined in concluding a series of agreements with all the other American republics. This further solidified our hemisphere against the common danger.

And then, a year ago, we launched, and are successfully carrying out, the largest armament production program we have ever undertaken.

We have added substantially to our splendid navy, and we have mustered our manpower to build up a new army which is already worthy of the highest traditions of our military service. . . .

We do not accept, we will not permit, this Nazi "shape of things to come." It will never be forced upon us, if we act in this present crisis with the wisdom and the courage which have distinguished our country in all the crises of the past.

Today, the Nazis have taken military possession of the greater part of Europe. In Africa they have occupied Tripoli and Libya, and they are threatening Egypt, the Suez Canal, and the Near East. But their plans do not stop there, for the Indian Ocean is the gateway to the farther East.

They also have the armed power at any moment to occupy Spain and Portugal; and that threat extends not only to French North Africa and the western end of the Mediterranean but it extends also to the Atlantic fortress of Dakar, and to the island outposts of the New World, the Azores and the Cape Verde Islands. . . .

Control or occupation by Nazi forces of any of the islands of the Atlantic would jeopardize the immediate safety of portions of North and South America, and of the island possessions of the United States, and, therefore, the ultimate safety of the continental United States itself.

Hitler's plan of world domination would be near its accomplishment today were it not for two factors. One is the epic resistance of Britain, her colonies, and the great dominions, fighting not only to maintain the existence of the island of Britain, but also to hold the Near East and Africa. The other is the magnificent defense of China, which will, I have reason to believe, increase in strength. All of these, together, are preventing the Axis from winning control of the seas by ships and aircraft.

The Axis powers can never achieve their objective of world domination unless they first obtain control of the seas. That is their supreme purpose today; and to achieve it, they must capture Great Britain. . . .

But if the Axis powers fail to gain control of the seas, then they are certainly defeated. Their dreams of world domination will then go by the board; and the criminal leaders who started this war will suffer inevitable disaster. . . .

All freedom—meaning freedom to live, and not freedom to conquer and subjugate other peoples—depends on freedom of the seas. All of American history—North, Central, and South American history—has been inevitably tied up with those words, "freedom of the seas." . . .

The blunt truth is this—and I reveal this with the full knowledge of the British government—the present rate of Nazi sinkings of merchant ships is more than three times as high as the capacity of British shipyards to replace them; it is more than twice the combined British and American output of merchant ships today.

We can answer this peril by two simultaneous measures: first, by speeding up and increasing our own great shipbuilding program; and second, by helping to cut down the losses on the high seas.

Attacks on shipping off the very shores of land which we are determined to protect present an actual military danger to the Americas. And that danger has recently been heavily underlined by the presence in Western Hemisphere waters of a Nazi battleship of great striking power. . . .

Defense today means more than merely fighting. It means morale, civilian as well as military; it means using every available resource; it means enlarging every useful plant. It means the use of a greater American common sense in discarding rumor and distorted statement. It means recognizing, for what they are, racketeers and fifth columnists, who are the incendiary bombs in this country of the moment. . . .

No one of us can waver for a moment in his courage or his faith.

We will not accept a Hitler-dominated world. And we will not accept a world, like the postwar world of the 1920s, in which the seeds of Hitlerism can again be planted and allowed to grow.

We will accept only a world consecrated to freedom of speech and expression—freedom of every person to worship God in his own way—freedom from want—and freedom from terror. . . .

As the president of a united and determined people, I say solemnly:

We reassert the ancient American doctrine of freedom of the seas.

We reassert the solidarity of the twenty-one American republics and the Dominion of Canada in the preservation of the independence of the hemisphere.

We have pledged material support to the other democracies of the world and we will fulfill that pledge.

We in the Americas will decide for ourselves whether, and when, and where, our American interests are attacked or our security is threatened.

We are placing our armed forces in strategic military position.

We will not hesitate to use our armed forces to repel attack.

We reassert our abiding faith in the vitality of our constitutional republic as a perpetual home of freedom, of tolerance, and of devotion to the word of God.

Therefore, with profound consciousness of my responsibilities to my countrymen and to my country's cause, I have tonight issued a proclamation that an unlimited national emergency exists and requires the strengthening of our defense to the extreme limit of our national power and authority.

The nation will expect all individuals and all groups to play their full parts, without stint, and without selfishness, and without doubt that our democracy will triumphantly survive.

I repeat the words of the signers of the Declaration of Independence—that little band of patriots, fighting long ago against overwhelming odds, but certain, as we are now, of ultimate victory: "With a firm reliance on the protection of Divine Providence, we mutually pledge to each other our lives, our fortunes, and our sacred honor."

In a message to Churchill in May 1941, Roosevelt, intending to encourage Churchill, actually depressed him: "You have fought a wholly justified delaying action and will continue to do so in other parts of Eastern Mediterranean, including North Africa and the Near East. Furthermore, . . . if additional withdrawals become necessary, they will all be part of the plan which at this stage of the war shortens British lines, greatly extends the Axis lines, and compels the enemy to expend great quantities of men and equipment. I am satisfied that both here and in Britain public opinion is growing to realize that even if you have to withdraw further in the Eastern Mediterranean, you will not allow any great debacle or surrender, and that in the last analysis the Naval control of the Indian Ocean and the Atlantic Ocean will in time win the war."

When he realized that his earlier message had the wrong effect, he wrote to Churchill again: "In my message of May 1, I did not intend to minimize in any degree the gravity of the situation, particularly as regards the Mediterranean. I am well aware of its strategic importance and I share your anxiety in regard to it."

Roosevelt promised that thirty ships would sail to the Middle East with supplies "within the next three weeks." The supplies would continue to be shipped "until there is a final decision in the Mediterranean. . . . My previous message merely meant to indicate that should the Mediterranean prove in the last analysis to be an impossible battleground, I do not feel that such fact alone would mean the defeat of our mutual interests. I say this because I believe the outcome of struggle is going to be decided in the Atlantic and unless Hitler can win there he cannot win anywhere in the world in the end."

Roosevelt used his September 11, 1941, Fireside Chat to delineate the precise state of the national defense with the American people.

❧ . . . IT IS TIME for all Americans, Americans of all the Americas, to stop being deluded by the romantic notion that the Americas can go on living happily and peacefully in a Nazi-dominated world.

Generation after generation, America has battled for the general policy of the freedom of the seas. And that policy is a very simple one, but a basic, a fundamental one. It means that no nation has the right to make the broad oceans of the world at great distances from the actual theater of land war unsafe for the commerce of others.

That has been our policy, proved time and time again, in all our history.

Our policy has applied from the earliest days of the Republic—and still applies not merely to the Atlantic but to the Pacific and to all other oceans as well.

Unrestricted submarine warfare in 1941 constitutes a defiance—an act of aggression—against that historic American policy.

It is now clear that Hitler has begun his campaign to

control the seas by ruthless force and by wiping out every vestige of international law, every vestige of humanity. . . .

The Nazi danger to our Western world has long ceased to be a mere possibility. The danger is here now—not only from a military enemy but from an enemy of all law, all liberty, all morality, all religion.

There has now come a time when you and I must see the cold, inexorable necessity of saying to these inhuman, unrestrained seekers of world conquest and permanent world domination by the sword: "You seek to throw our children and our children's children into your form of terrorism and slavery. You have now attacked our own safety. You shall go no further." . . .

My obligation as president is historic; it is clear. It is inescapable.

It is no act of war on our part when we decide to protect the seas that are vital to American defense. The aggression is not ours. Ours is solely defense.

But let this warning be clear. From now on, if German or Italian vessels of war enter the waters, the protection of which is necessary for American defense, they do so at their own peril.

The orders which I have given as commander in chief of the United States Army and Navy are to carry out that policy at once.

The sole responsibility rests upon Germany. There will be no shooting unless Germany continues to seek it. . . .

Roosevelt was also worried that the shooting might come from across the Pacific. On November 24, 1941, FDR cabled Churchill the terms that the U.S. was prepared to offer Japan.

In a personal postscript, Roosevelt wrote, "This seems to me a fair proposition for the Japanese, but its acceptance or rejection is really a matter of internal Japanese politics. I am not very hopeful, and we must be prepared for real trouble, possibly soon."

Two weeks later, the effort would prove to be in vain. The day after Pearl Harbor was bombed, in one of the most famous speeches in American history, Roosevelt addressed Congress to request a declaration of war between Japan and the United States.

MR. VICE PRESIDENT, Mr. Speaker, and Members of the Senate and House of Representatives: Yesterday, December 7, 1941—a date which will live in infamy—the United States of America was suddenly and deliberately attacked by naval and air forces of the Empire of Japan.

The United States was at peace with that nation and, at the solicitation of Japan, was still in conversation with its government and its emperor looking toward the maintenance of peace in the Pacific. Indeed, one hour after Japanese air squadrons had commenced bombing in the American island of Oahu, the Japanese ambassador to the United States and his colleague delivered to our secretary of state a formal reply to a recent American message. And while this reply stated that it seemed useless to continue the existing diplomatic negotiations, it contained no threat or hint of war or of armed attack.

It will be recorded that the distance of Hawaii from Japan makes it obvious that the attack was deliberately planned many days or even weeks ago. During the intervening time the Japanese government has deliberately sought to deceive the United States by false statements and expressions of hope for continued peace.

The attack yesterday on the Hawaiian Islands has caused severe damage to American naval and military forces. I regret to tell you that very many American lives have been lost. In addition American ships have been reported torpedoed on the high seas between San Francisco and Honolulu.

Yesterday the Japanese government also launched an attack against Malaya.

Last night Japanese forces attacked Hong Kong.

Last night Japanese forces attacked Guam.

Last night Japanese forces attacked the Philippine Islands.

Last night the Japanese attacked Wake Island.

And this morning the Japanese attacked Midway Island.

Japan has, therefore, undertaken a surprise offensive extending throughout the Pacific area. The facts of yesterday and today speak for themselves. The people of the United States have already formed their opinions and well understand the implications to the very life and safety of our nation.

As commander in chief of the army and navy I have directed that all measures be taken for our defense.

But always will our whole nation remember the character of the onslaught against us.

No matter how long it may take us to overcome this premeditated invasion, the American people in their righteous might will win through to absolute victory.

I believe that I interpret the will of the Congress and of the people when I assert that we will not only defend ourselves to the uttermost but will make it very certain that this form of treachery shall never again endanger us.

Hostilities exist. There is no blinking at the fact that our people, our territory, and our interests are in grave danger.

With confidence in our armed forces, with the unbounding determination of our people, we will gain the inevitable triumph—so help us God.

I ask that the Congress declare that since the unprovoked and dastardly attack by Japan on Sunday, December 7, 1941, a state of war has existed between the United States and the Japanese Empire.

At the time, he also said, "No matter what our enemies . . . in their desperation may attempt to do to us—we will say as the people of London have said, 'We can take it,' and what is more, we can give it back—with compound interest."

In his December 9, 1941, Fireside Chat, he tried to get the nation ready for world war.

THE SUDDEN CRIMINAL attacks perpetrated by the Japanese in the Pacific provide the climax of a decade of international immorality.

Powerful and resourceful gangsters have banded together to make war upon the whole human race. Their challenge has now been flung at the United States of America. The Japanese have treacherously violated the long-standing peace between us. Many American soldiers and sailors have been killed by enemy action. American ships have been sunk; American airplanes have been destroyed.

The Congress and the people of the United States have accepted that challenge.

Together with other free peoples, we are now fighting to maintain our right to live among our world neighbors in freedom and in common decency, without fear of assault.

I have prepared the full record of our past relations with Japan, and it will be submitted to the Congress. It begins with the visit of Commodore Perry to Japan eighty-eight years ago. It ends with the visit of two Japanese emissaries to the secretary of state last Sunday, an hour after Japanese forces had loosed their bombs and machine guns against our flag, our forces, and our citizens.

I can say with utmost confidence that no Americans, today or a thousand years hence, need feel anything but pride in our patience and in our efforts through all the years toward achieving a peace in the Pacific which would be fair and honorable to every nation, large or small. And no honest person, today or a thousand years hence, will be able to suppress a sense of indignation and horror at the treachery committed by the military dictators of Japan, under the very shadow of the flag of peace borne by their special envoys in our midst.

The course that Japan has followed for the past ten years in Asia has paralleled the course of Hitler and Mussolini in Europe and in Africa. Today, it has become far more than a parallel. It is actual collaboration so well calculated that all the continents of the world, and all the oceans, are now considered by the Axis strategists as one gigantic battlefield.

In 1931, ten years ago, Japan invaded Manchukuo—without warning.

In 1935, Italy invaded Ethiopia—without warning.

In 1938, Hitler occupied Austria—without warning.

In 1939, Hitler invaded Czechoslovakia—without warning.

Later in 1939, Hitler invaded Poland—without warning.

In 1940, Hitler invaded Norway, Denmark, the Netherlands, Belgium, and Luxembourg—without warning.

In 1940, Italy attacked France and later Greece—without warning.

And this year, in 1941, the Axis powers attacked Yugoslavia and Greece and they dominated the Balkans—without warning.

In 1941, also, Hitler invaded Russia—without warning.

And now Japan has attacked Malaya and Thailand and the United States—without warning.

It is all of one pattern.

We are now in this war. We are all in it—all the way. Every single man, woman, and child is a partner in the most tremendous undertaking of our American history. We must share together the bad news and the good news, the defeats and the victories—the changing fortunes of war.

So far, the news has been all bad. We have suffered a serious setback in Hawaii. Our forces in the Philippines, which include the brave people of that commonwealth, are taking punishment, but are defending themselves vigorously. The reports from Guam and Wake and Midway islands are still confused, but we must be prepared for the announcement that all these three outposts have been seized.

The casualty lists of these first few days will undoubtedly be large. I deeply feel the anxiety of all of the families of the men in our armed forces and the relatives of people in cities which have been bombed. I can only give them my solemn promise that they will get news just as quickly as possible.

This government will put its trust in the stamina of the American people, and will give the facts to the public just as soon as two conditions have been fulfilled: first, that the information has been definitely and officially confirmed; and, second, that the release of the information at the time it is received will not prove valuable to the enemy directly or indirectly.

Most earnestly I urge my countrymen to reject all rumors. These ugly little hints of complete disaster fly thick and fast in wartime. They have to be examined and appraised.

As an example, I can tell you frankly that until further surveys are made, I have not sufficient information to state the exact damage which has been done to our naval vessels at Pearl Harbor. Admittedly the damage is serious. But no one can say how serious, until we know how much of this damage can be repaired and how quickly the necessary repairs can be made.

I cite as another example a statement made on Sunday night that a Japanese carrier had been located and sunk off the Canal Zone. And when you hear statements that are attributed to what they call "an authoritative source," you can be reasonably sure from now on that under these war circumstances the "authoritative source" is not any person in authority.

Many rumors and reports which we now hear originate with enemy sources. For instance, today the Japanese are claiming that as a result of their one action against Hawaii they have gained naval supremacy in the Pacific. This is an old trick of propaganda which has been used innumerable times by the Nazis. The purposes of such fantastic claims are, of course, to spread fear and confusion among us, and to goad us into revealing military information which our enemies are desperately anxious to obtain.

Our government will not be caught in this obvious trap—and neither will the people of the United States.

It must be remembered by each and every one of us that our free and rapid communication these days must be greatly restricted in wartime. It is not possible to receive full, speedy, accurate reports from distant areas of combat. This is particularly true where naval operations are concerned. For in these days of the marvels of radio it is often impossible for the commanders of various units to report their activities by radio at all, for the very simple reason that this information would become available to the enemy, and would disclose their position and their plan of defense or attack.

Of necessity there will be delays in officially confirming or denying reports of operations, but we will not hide facts from the country if we know the facts and if the enemy will not be aided by their disclosure.

To all newspapers and radio stations—all those who reach the eyes and ears of the American people—I say this: You have a most grave responsibility to the nation now and for the duration of this war.

If you feel that your government is not disclosing enough of the truth, you have every right to say so. But—in the absence of all the facts, as revealed by official sources—you have no right in the ethics of patriotism to deal out unconfirmed reports in such a way as to make people believe that they are gospel truth.

Every citizen, in every walk of life, shares this same responsibility. The lives of our soldiers and sailors—the whole future of this nation—depend upon the manner in which each and every one of us fulfills his obligation to our country.

Now a word about the recent past—and the future. A year and a half has elapsed since the fall of France, when the whole world first realized the mechanized might which the Axis nations had been building for so many years. America has used that year and a half to great advantage. Knowing that the attack might reach us in all too short a time, we immediately began greatly to increase our industrial strength and our capacity to meet the demands of modern warfare.

Precious months were gained by sending vast quantities of our war material to the nations of the world still able to resist Axis aggression. Our policy rested on the fundamental truth that the defense of any country resisting Hitler or Japan was in the long run the defense of our own country. That policy has been justified. It has given us time, invaluable time, to build our American assembly lines of production.

Assembly lines are now in operation. Others are being rushed to completion. A steady stream of tanks and planes, of guns and ships, and shells and equipment—that is what these eighteen months have given us.

But it is all only a beginning of what still has to be done. We must be set to face a long war against crafty and powerful bandits. The attack at Pearl Harbor can be repeated at any one of many points, points in both oceans and along both our coast lines and against all the rest of the hemisphere.

It will not only be a long war, it will be a hard war. That is the basis on which we now lay all our plans. That is the yardstick by which we measure what we shall need and demand; money, materials, doubled and quadrupled production—ever-increasing. The production must be not only for our own army and navy and air forces. It must reinforce the other armies and navies and air forces fighting the Nazis and the warlords of Japan throughout the Americas and throughout the world.

I have been working today on the subject of production. Your government has decided on two broad policies.

The first is to speed up all existing production by working on a seven-day week basis in every war industry, including the production of essential raw materials.

The second policy, now being put into form, is to rush additions to the capacity of production by building more new plants, by adding to old plants, and by using the many smaller plants for war needs.

Over the hard road of the past months, we have at times met obstacles and difficulties, divisions and disputes, indifference and callousness. That is now all past—and, I am sure, forgotten.

The fact is that the country now has an organization in Washington built around men and women who are recognized experts in their own fields. I think the country knows that the people who are actually responsible in each and every one of these many fields are pulling together with a teamwork that has never before been excelled.

On the road ahead there lies hard work—grueling work—day and night, every hour and every minute.

I was about to add that ahead there lies sacrifice for all of us.

But it is not correct to use that word. The United States does not consider it a sacrifice to do all one can, to give one's best to our nation, when the nation is fighting for its existence and its future life.

It is not a sacrifice for any man, old or young, to be in the army or the navy of the United States. Rather is it a privilege.

It is not a sacrifice for the industrialist or the wage earner, the farmer or the shopkeeper, the trainman or the doctor, to pay more taxes, to buy more bonds, to forego extra profits, to work longer or harder at the task for which he is best fitted. Rather it is a privilege.

It is not a sacrifice to do without many things to which we are accustomed if the national defense calls for doing without.

A review this morning leads me to the conclusion that at present we shall not have to curtail the normal use of articles of food. There is enough food today for all of us and enough left over to send to those who are fighting on the same side with us.

But there will be a clear and definite shortage of metals of many kinds for civilian use, for the very good reason that in our increased program we shall need for war purposes more than half of that portion of the principal metals which during the past year have gone into articles for civilian use. Yes, we shall have to give up many things entirely.

And I am sure that the people in every part of the nation are prepared in their individual living to win this war. I am sure that they will cheerfully help to pay a large part of its financial cost while it goes on. I am sure they will cheerfully give up those material things that they are asked to give up.

And I am sure that they will retain all those great spiritual things without which we cannot win through.

I repeat that the United States can accept no result save victory, final and complete. Not only must the shame of Japanese treachery be wiped out, but the sources of international brutality, wherever they exist, must be absolutely and finally broken.

In my message to the Congress yesterday I said that we "will make it very certain that this form of treachery shall never again endanger us." In order to achieve that certainty, we must begin the great task that is before us by abandoning once and for all the illusion that we can ever again isolate ourselves from the rest of humanity.

In these past few years and, most violently, in the past three days we have learned a terrible lesson.

It is our obligation to our dead—it is our sacred obligation to their children and to our children—that we must never forget what we have learned.

And what we all have learned is this:

There is no such thing as security for any nation—or any individual—in a world ruled by the principles of gangsterism.

There is no such thing as impregnable defense against powerful aggressors who sneak up in the dark and strike without warning.

We have learned that our ocean-girt hemisphere is not immune from severe attack—that we cannot measure our safety in terms of miles on any map any more.

We may acknowledge that our enemies have performed a brilliant feat of deception, perfectly timed and executed with great skill. It was a thoroughly dishonorable deed, but we must face the fact that modern warfare as conducted in the Nazi manner is a dirty business. We don't like it—we didn't want to get into it—but we are in it and we're going to fight it with everything we've got.

I do not think any American has any doubt of our ability to administer proper punishment to the perpetrators of these crimes.

Your government knows that for weeks Germany has been telling Japan that if Japan did not attack the United States, Japan would not share in dividing the spoils with Germany when peace came. She was promised by Germany that if she came in she would receive the complete and perpetual control of the whole of the Pacific area—and that means not only the Far East, but also all of the islands in the Pacific, and also a stranglehold on the west coast of North, Central, and South America.

We know also that Germany and Japan are conducting their military and naval operations in accordance with a joint plan. That plan considers all peoples and nations which are not helping the Axis powers as common enemies of each and every one of the Axis powers.

That is their simple and obvious grand strategy. And that is why the American people must realize that it can be matched only with similar grand strategy. We must realize for example that Japanese successes against the United States in the Pacific are helpful to German operations in Libya; that any German success against the Caucasus is inevitably an assistance to Japan in her operations against the Dutch East Indies; that a German attack against Algiers or Morocco opens the way to a German attack against South America, and the Canal.

On the other side of the picture, we must learn also to know that guerrilla warfare against the Germans in, let us say, Serbia or Norway helps us; that a successful Russian offensive against the Germans helps us; and that British successes on land or sea in any part of the world strengthen our hands.

Remember always that Germany and Italy, regardless of any formal declaration of war, consider themselves at war with the United States at this moment just as much as they consider themselves at war with Britain or Russia. And Germany puts all the other republics of the Americas into the same category of enemies. The people of our sister republics of this hemisphere can be honored by that fact.

The true goal we seek is far above and beyond the ugly field of battle. When we resort to force, as now we must, we are determined that this force shall be directed toward ultimate good as well as against immediate evil. We Americans are not destroyers—we are builders.

We are now in the midst of a war, not for conquest, not for vengeance, but for a world in which this nation, and all that this nation represents, will be safe for our children. We expect to eliminate the danger from Japan, but it would serve us ill if we accomplished that and found that the rest of the world was dominated by Hitler and Mussolini.

We are going to win the war and we are going to win the peace that follows.

And in the difficult hours of this day—through dark days that may be yet to come—we will know that the vast majority of the members of the human race are on our side. Many of them are fighting with us. All of them are praying for us. For in representing our cause, we represent theirs as well—our hope and their hope for liberty under God.

FDR told his speechwriter, Samuel Rosenman, "I want to explain the war in laymen's language; if they understand the problem and what we are driving at, I am sure that they can take any kind of bad news right on the chin."

Within days, the war was joined. On December 11, 1941, Roosevelt sent the inevitable message to Congress requesting a declaration of war between Germany and Italy and the United States.

ON THE MORNING of December 11, the government of Germany, pursuing its course of world conquest, declared war against the United States.

The long known and the long expected has thus taken place. The forces endeavoring to enslave the entire world now are moving toward this hemisphere.

Never before has there been a greater challenge to life, liberty, and civilization.

Delay invites greater danger. Rapid and united effort by all of the peoples of the world who are determined to remain free will insure a world victory of the forces of justice and of righteousness over the forces of savagery and of barbarism.

Italy also has declared war against the United States.

I therefore request the Congress to recognize a state of war between the United States and Germany, and between the United States and Italy.

Only a month later, on January 6, 1942, Roosevelt gave a remarkably upbeat assessment of the state of the union in his ninth annual message to Congress.

IN FULFILLING MY duty to report upon the state of the Union, I am proud to say to you that the spirit of the American people was never higher than it is today—the Union was never more closely knit together—this country was never more deeply determined to face the solemn tasks before it.

The response of the American people has been instantaneous, and it will be sustained until our security is assured.

Exactly one year ago today I said to this Congress: "When the dictators . . . are ready to make war upon us, they will not wait for an act of war on our part. . . . They—not we—will choose the time and the place and the method of their attack."

We now know their choice of the time: a peaceful Sunday morning, December 7, 1941.

We know their choice of the place: an American outpost in the Pacific.

We know their choice of the method: the method of Hitler himself.

Japan's scheme of conquest goes back half a century. It was not merely a policy of seeking living room—it was a plan which included the subjugation of all the peoples in the Far East and in the islands of the Pacific, and the domination of that ocean by Japanese military and naval control of the western coasts of North, Central, and South America.

The development of this ambitious conspiracy was marked by the war against China in 1894; the subsequent occupation of Korea; the war against Russia in 1904; the illegal fortification of the mandated Pacific islands following 1920; the seizure of Manchuria in 1931; and the invasion of China in 1937.

A similar policy of criminal conquest was adopted by Italy. The Fascists first revealed their imperial designs in Libya and Tripoli. In 1935 they seized Abyssinia. Their goal was the domination of all North Africa, Egypt, parts of France, and the entire Mediterranean world.

But the dreams of empire of the Japanese and Fascist leaders were modest in comparison with the gargantuan aspirations of Hitler and his Nazis. Even before they came to power in 1933, their plans for that conquest had been drawn. Those plans provided for ultimate domination, not of any one section of the world, but of the whole earth and all the oceans on it.

When Hitler organized his Berlin-Rome-Tokyo alliance, all these plans of conquest became a single plan. Under this, in addition to her own schemes of conquest, Japan's role was obviously to cut off our supply of weapons of war to Britain and Russia and China—weapons which increasingly were speeding the day of Hitler's doom. The act of Japan at Pearl Harbor was intended to stun us—to terrify us to such an extent that we would divert our industrial and military strength to the Pacific area, or even to our own continental defense.

The plan has failed in its purpose. We have not been stunned. We have not been terrified or confused. This very reassembling of the Seventy-seventh Congress today is proof of that; for the mood of quiet, grim resolution which here prevails bodes ill for those who conspired and collaborated to murder world peace.

That mood is stronger than any mere desire for revenge. It expresses the will of the American people to make very certain that the world will never so suffer again.

Admittedly, we have been faced with hard choices. It

was bitter, for example, not to be able to relieve the heroic and historic defenders of Wake Island. It was bitter for us not to be able to land a million men in a thousand ships in the Philippine Islands.

But this adds only to our determination to see to it that the Stars and Stripes will fly again over Wake and Guam. Yes, see to it that the brave people of the Philippines will be rid of Japanese imperialism, and will live in freedom, security, and independence.

Powerful and offensive actions must and will be taken in proper time. The consolidation of the United Nations' total war effort against our common enemies is being achieved.

That was and is the purpose of conferences which have been held during the past two weeks in Washington and Moscow and Chungking. That is the primary objective of the declaration of solidarity signed in Washington on January 1, 1942, by twenty-six nations united against the Axis powers.

Plans have been laid here and in the other capitals for coordinated and cooperative action by all the United Nations—military action and economic action. Already we have established, as you know, unified command of land, sea, and air forces in the southwestern Pacific theater of war. There will be a continuation of conferences and consultations among military staffs, so that the plans and operations of each will fit into the general strategy designed to crush the enemy. We shall not fight isolated wars—each nation going its own way. These twenty-six nations are united—not in spirit and determination alone, but in the broad conduct of the war in all its phases.

For the first time since the Japanese and the Fascists and the Nazis started along their blood-stained course of conquest they now face the fact that superior forces are assembling against them. Gone forever are the days when the aggressors could attack and destroy their victims one by one without unity of resistance. We of the United Nations will so dispose our forces that we can strike at the common enemy wherever the greatest damage can be done him.

The militarists of Berlin and Tokyo started this war. But the massed, angered forces of common humanity will finish it.

Destruction of the material and spiritual centers of civilization—this has been and still is the purpose of Hitler and his Italian and Japanese chessmen. They would wreck the power of the British Commonwealth and Russia and China and the Netherlands—and then combine all their forces to achieve their ultimate goal, the conquest of the United States.

They know that victory for us means victory for freedom.

They know that victory for us means victory for the institution of democracy—the ideal of the family, the simple principles of common decency and humanity.

They know that victory for us means victory for religion.

And they could not tolerate that. The world is too small to provide adequate "living room" for both Hitler and God. In proof of that, the Nazis have now announced their plan for enforcing their new German, pagan religion all over the world—a plan by which the Holy Bible and the Cross of Mercy would be displaced by *Mein Kampf* and the swastika and the naked sword.

Our own objectives are clear; the objective of smashing the militarism imposed by warlords upon their enslaved peoples—the objective of liberating the subjugated nations—the objective of establishing and securing freedom of speech, freedom of religion, freedom from want, and freedom from fear everywhere in the world.

We shall not stop short of these objectives, nor shall we be satisfied merely to gain them and then call it a day. I know that I speak for the American people—and I have good reason to believe that I speak also for all the other peoples who fight with us—when I say that this time we are determined not only to win the war, but also to maintain the security of the peace that will follow. . . .

The superiority of the United Nations in munitions and ships must be overwhelming—so overwhelming that the Axis nations can never hope to catch up with it. And so, in order to attain this overwhelming superiority the United States must build planes and tanks and guns and ships to the utmost limit of our national capacity. We have the ability and capacity to produce arms not only for our own forces, but also for the armies, navies, and air forces fighting on our side.

And our overwhelming superiority of armament must be adequate to put weapons of war at the proper time into the hands of those men in the conquered nations who stand ready to seize the first opportunity to revolt against their German and Japanese oppressors, and against the traitors in their own ranks, known by the already infamous name of Quislings. And I think that it is a fair prophecy to say that, as we get guns to the patriots in those lands, they too will fire shots heard 'round the world.

This production of ours in the United States must be raised far above present levels, even though it will mean the dislocation of the lives and occupations of millions of our own people. We must raise our sights all along the production line. Let no man say it cannot be done. It must be done—and we have undertaken to do it. . . .

Our task is hard—our task is unprecedented and the time is short. We must strain every existing armament-producing facility to the utmost. We must convert every available plant and tool to war production. That goes all the way from the greatest plants to the smallest—from the huge automobile industry to the village machine shop.

Production for war is based on men and women—the human hands and brains which collectively we call Labor. Our workers stand ready to work long hours; to turn out more in a day's work; to keep the wheels turning and the fires burning twenty-four hours a day, and seven days a week. They realize well that on the speed and efficiency of their work depend the lives of their sons and their brothers on the fighting fronts.

Production for war is based on metals and raw materials—steel, copper, rubber, aluminum, zinc, tin. Greater and greater quantities of them will have to be diverted to war purposes. Civilian use of them will have to be cut further and still further—and, in many cases, completely eliminated.

War costs money. So far, we have hardly even begun to pay for it. We have devoted only 15 percent of our national income to national defense. As will appear in my Budget Message tomorrow, our war program for the coming fiscal year will cost $56 billion or, in other words, more than half of the estimated annual national income. That means taxes and bonds and bonds and taxes. It means cutting luxuries and other nonessentials. In a word, it means an "all-out" war by individual effort and family effort in a united country.

Only this all-out scale of production will hasten the ultimate all-out victory. Speed will count. Lost ground can always be regained—lost time never. Speed will save lives; speed will save this nation which is in peril; speed will save our freedom and our civilization and slowness has never been an American characteristic.

We cannot wage this war in a defensive spirit. As our power and our resources are fully mobilized, we shall carry the attack against the enemy—we shall hit him and hit him again wherever and whenever we can reach him.

We must keep him far from our shores, for we intend to bring this battle to him on his own home grounds.

American armed forces must be used at any place in all the world where it seems advisable to engage the forces of the enemy. In some cases these operations will be defensive, in order to protect key positions. In other cases, these operations will be offensive, in order to strike at the common enemy, with a view to his complete encirclement and eventual total defeat.

American armed forces will operate at many points in the Far East.

American armed forces will be on all the oceans helping to guard the essential communications which are vital to the United Nations.

American land and air and sea forces will take stations in the British Isles—which constitute an essential fortress in this great world struggle.

American armed forces will help to protect this hemisphere and also help to protect bases outside this hemisphere, which could be used for an attack on the Americas.

If any of our enemies, from Europe or from Asia, attempt long-range raids by "suicide" squadrons of bombing planes, they will do so only in the hope of terrorizing our people and disrupting our morale. Our people are not afraid of that. We know that we may have to pay a heavy price for freedom. We will pay this price with a will. Whatever the price, it is a thousand times worth it. No matter what our enemies, in their desperation, may attempt to do to us—we will say, as the people of London have said, "We can take it." And what's more we can give it back—and we will give it back—with compound interest.

Many people ask, "When will this war end?" There is only one answer to that. It will end just as soon as we make it end, by our combined efforts, our combined strength, our combined determination to fight through and work through until the end—the end of militarism in Germany and Italy and Japan. Most certainly we shall not settle for less.

That is the spirit in which discussions have been conducted during the visit of the British prime minister to Washington. Mr. Churchill and I understand each other, our motives, and our purposes. Together, during the past two weeks, we have faced squarely the major military and economic problems of this greatest world war.

All in our nation have been cheered by Mr. Churchill's visit. We have been deeply stirred by his great message to us. He is welcome in our midst, and we unite in wishing him a safe return to his home.

For we are fighting on the same side with the British people, who fought alone for long, terrible months, and withstood the enemy with fortitude and tenacity and skill.

We are fighting on the same side with the Russian people who have seen the Nazi hordes swarm up to the very gates of Moscow, and who with almost superhuman will and courage have forced the invaders back into retreat.

We are fighting on the same side as the brave people of China—those millions who for four and a half long years have withstood bombs and starvation and have whipped the invaders time and again in spite of the superior Japanese equipment and arms.

Yes, we are fighting on the same side as the indomitable Dutch.

We are fighting on the same side as all the other governments in exile, whom Hitler and all his armies and all his Gestapo have not been able to conquer.

But we of the United Nations are not making all this sacrifice of human effort and human lives to return to the kind of world we had after the last world war.

We are fighting today for security, for progress, and for peace, not only for ourselves but for all men, not only for one generation but for all generations. We are fighting to cleanse the world of ancient evils, ancient ills.

Our enemies are guided by brutal cynicism, by unholy contempt for the human race. We are inspired by a faith that goes back through all the years to the first chapter of the book of Genesis: "God created man in His own image."

We on our side are striving to be true to that divine heritage. We are fighting, as our fathers have fought, to uphold the doctrine that all men are equal in the sight of God. Those on the other side are striving to destroy this deep belief and to create a world in their own image—a world of tyranny and cruelty and serfdom.

That is the conflict that day and night now pervades our lives. No compromise can end that conflict. There never has been—there never can be—successful compromise between good and evil. Only total victory can reward the champions of tolerance, and decency, and freedom, and faith.

But total victory has its costs, and as the war dragged on, Roosevelt sent the following message to Congress on September 17, 1943.

. . . THE OUTPUT OF underwater ordnance (torpedoes, mines, and depth charges) during the first half of 1943 was equal to the total production of 1942.

During the month of August 1943, we produced almost as many torpedoes as during all of World War I.

Anyone who has had to build a single factory, tool it up, get the necessary help, set up an assembly line, produce and ship the product will have some idea of what that amount of production has meant.

We have had to raise and equip armed forces approaching ten million men. Simultaneously, in spite of this drain on our manpower, we have had to find millions more men and millions of women to operate our war factories, arsenals, shipyards, essential civilian industries—and the farms and mines of America.

There have been the problems of increasing greatly the output of our natural resources not only for our own army and navy and for our civilians at home, but also for our allies and our own forces all over the world.

Since the outbreak of war in Europe, we have increased our output of petroleum by 66 percent. We have stepped up our bituminous coal production by 40 percent;

chemicals by 300 percent; iron ore by 125 percent; hydroelectric power by 79 percent; and steel by 106 percent.

There were the problems of raising and distributing more food than ever before in our history for our armed services, for our own people, and to help feed our allies.

There was the formidable problem of establishing a rationing system of the necessities of life which would be fair to all of our people.

There was the difficulty of keeping prices from skyrocketing and fighting off the serious specter of inflation.

There was the problem of transporting millions of men and hundreds of millions of tons of weapons and supplies all over our own country and also to all corners of the world. This necessitated the largest railroad and shipping operations in all history.

There were the problems involved in our vast purchases in foreign countries; in our control of foreign funds, located in this country; in our custody of alien property; in our occupation of liberated areas. There were new problems of communications, of censorship, of war information.

There was the problem of maintaining proper management-labor relations; of fair treatment and just compensation to our millions of war workers; of avoiding strikes; of preventing the exploitation of workers or natural resources by those who would seek to become war profiteers and war millionaires.

There were the problems of civilian defense, of lend-lease, of subcontracting war contracts to smaller businesses, of building up stockpiles of strategic material whose normal sources have been seized by the enemy, such as rubber and tin.

There was the problem of providing housing for millions of new war workers all over the country.

And touching all of these, there was the great problem of raising the money to pay for all of them.

No sincere, sensible person doubts that in such an unprecedented, breathtaking enterprise errors of honest judgment were bound to creep in, and that occasional disputes among conscientious officials were bound to occur. And if anyone thinks that we, working under our democratic system, have made major mistakes in this war, he should take a look at some of the blunders made by our enemies in the so-called "efficient" dictatorships. . . .

As the war grows tougher and as new problems constantly arise in our domestic economy, changes in methods and changes in legislation may become necessary.

We should move for the greater economic protection of our returning men and women in the armed forces and for greater educational opportunities for them. And for all our citizens we should provide a further measure of social security in order to protect them against certain continuing hazards of life.

All these things, as well as eventual demobilization, should be studied now and much of the necessary legislation should be enacted. I do not mean that this statement should be regarded in any way as an intimation that we are approaching the end of the war. Such an intimation could not be based either on fact or on reason. But when the war ends, we do not want to be caught again without planning or legislation, such as occurred at the end of the last war. . . .

Finally, as the war progresses, we seek a national cooperation with other nations toward the end that world aggression be ended and that fair international relationships be established on a permanent basis. The policy of the Good Neighbor has shown such success in the hemisphere of the Americas that its extension to the whole world seems to be the logical next step. In that way we can begin to keep faith with our sons and daughters who are fighting for freedom and justice and security at home and abroad.

That source for "national cooperation" would manifest itself in a meeting between FDR, Stalin and Churchill in Teheran in November 1943. Roosevelt's Fireside Chat on Christmas Eve that year, in which he discussed the event, provided America with reasons to look forward to the coming year.

❧ I HAVE RECENTLY returned from extensive journeyings in the region of the Mediterranean and as far as the borders of Russia. I have conferred with the leaders of Britain and Russia and China on military matters of the present—especially on plans for stepping up our successful attack on our enemies as quickly as possible and from many different points of the compass.

On this Christmas Eve there are over 10 million men in the armed forces of the United States alone. One year ago 1.7 million were serving overseas. Today, this figure has been more than doubled to 3.8 million on duty overseas. By next July 1, that number overseas will rise to over 5 million men and women.

That this is truly a world war was demonstrated to me when arrangements were being made with our overseas broadcasting agencies for the time to speak today to our soldiers, sailors, marines, and merchant seamen in every part of the world. In fixing the time for this broadcast, we took into consideration that at this moment here in the United States, and in the Caribbean and on the northeast coast of South America, it is afternoon. In Alaska and Hawaii and the mid-Pacific, it is still morning. In Iceland, in Great Britain, in North Africa, in Italy and the Middle East, it is now evening.

In the Southwest Pacific, in Australia, in China and Burma and India, it is already Christmas Day. So we can correctly say that at this moment, in those Far Eastern parts where Americans are fighting, today is tomorrow.

But everywhere throughout the world—throughout this war that covers the world—there is a special spirit that has warmed our hearts since our earliest childhood, a spirit that brings us close to our homes, our families, our friends and neighbors—the Christmas spirit of "peace on earth, goodwill toward men." It is an unquenchable spirit.

During the past years of international gangsterism and brutal aggression in Europe and in Asia, our Christmas celebrations have been darkened with apprehension for the future. We have said, "Merry Christmas, Happy New Year," but we have known in our hearts that the clouds which have hung over our world have prevented us from saying it with full sincerity and conviction.

And even this year, we still have much to face in the way of further suffering, and sacrifice, and personal tragedy. Our men, who have been through the fierce battles in the Solomons, the Gilberts, Tunisia, and Italy, know from their own experience and knowledge of modern war that many bigger and costlier battles are still to be fought.

But—on Christmas Eve this year—I can say to you that at last we may look forward into the future with real, substantial confidence that, however great the cost, "peace on earth, goodwill toward men" can be and will be realized and insured. This year I can say that. Last year I could *not* do more than express a hope. Today I express a certainty though the cost may be high and the time may be long.

Within the past year—within the past few weeks—history has been made, and it is far better history for the whole human race than any that we have known, or even dared to hope for, in these tragic times through which we pass.

A great beginning was made in the Moscow Conference last October by Mr. Molotov, Mr. Eden, and our own Mr. Hull. There and then the way was paved for the later meetings.

At Cairo and Teheran we devoted ourselves not only to military matters; we devoted ourselves also to consideration of the future—to plans for the kind of world which alone can justify all the sacrifices of this war.

Of course, as you all know, Mr. Churchill and I have happily met many times before, and we know and understand each other very well. Indeed, Mr. Churchill has become known and beloved by many millions of Americans, and the heartfelt prayers of all of us have been with this great citizen of the world in his recent serious illness.

The Cairo and Teheran conferences, however, gave me my first opportunity to meet the Generalissimo, Chiang Kai-shek, and Marshal Stalin—and to sit down at the table with these unconquerable men and talk with them face to face. We had planned to talk to each other across the table at Cairo and Teheran; but we soon found that we were all on the same side of the table. We came to the conferences with faith in each other. But we needed the personal contact. And now we have supplemented faith with definite knowledge.

It was well worth traveling thousands of miles over land and sea to bring about this personal meeting, and to gain the heartening assurance that we are absolutely agreed with one another on all the major objectives—and on the military means of attaining them.

At Cairo, Prime Minister Churchill and I spent four days with the Generalissimo, Chiang Kai-shek. It was the first time that we had an opportunity to go over the complex situation in the Far East with him personally. We were able not only to settle upon definite military strategy, but also to discuss certain long-range principles which we believe can assure peace in the Far East for many generations to come.

Those principles are as simple as they are fundamental. They involve the restoration of stolen property to its rightful owners, and the recognition of the rights of millions of people in the Far East to build up their own forms of self-government without molestation. Essential to all peace and security in the Pacific and the rest of the world is the permanent elimination of the Empire of Japan as a potential force of aggression. Never again must our soldiers and sailors and marines—and other soldiers, sailors, and marines—be compelled to fight from island to island as they are fighting so gallantly and so successfully today.

Increasingly powerful forces are now hammering at the Japanese at many points over an enormous arc which curves down through the Pacific from the Aleutians to the jungles of Burma. Our own army and navy, our air forces, the Australians and New Zealanders, the Dutch, and the British land, air, and sea forces are all forming a band of steel which is slowly but surely closing in on Japan.

On the mainland of Asia, under the Generalissimo's leadership, the Chinese ground and air forces augmented by American air forces are playing a vital part in starting the drive which will push the invaders into the sea.

Following out the military decisions at Cairo, General Marshall has just flown around the world and has had conferences with General MacArthur and Admiral Nimitz—conferences which will spell plenty of bad news for the Japs in the not too far distant future.

I met in the Generalissimo a man of great vision, great courage, and a remarkably keen understanding of the problems of today and tomorrow. We discussed all the manifold military plans for striking at Japan with decisive force from many directions, and I believe I can say that he returned to Chungking with the positive assurance of total victory over our common enemy. Today we and the Republic of China are closer together than ever before in deep friendship and in unity of purpose.

After the Cairo conference, Mr. Churchill and I went by airplane to Teheran. There we met with Marshal Stalin. We talked with complete frankness on every conceivable subject connected with the winning of the war and the establishment of a durable peace after the war.

Within three days of intense and consistently amicable discussions, we agreed on every point concerned with the launching of a gigantic attack upon Germany.

The Russian army will continue its stern offensives on Germany's eastern front, the Allied armies in Italy and Africa will bring relentless pressure on Germany from the south, and now the encirclement will be complete as great American and British forces attack from other points of the compass.

The commander selected to lead the combined attack

from these other points is General Dwight D. Eisenhower. His performances in Africa, in Sicily, and in Italy have been brilliant. He knows by practical and successful experience the way to coordinate air, sea, and land power. All of these will be under his control. Lieutenant General Carl D. Spaatz will command the entire American strategic bombing force operating against Germany.

General Eisenhower gives up his command in the Mediterranean to a British officer whose name is being announced by Mr. Churchill. We now pledge that new commander that our powerful ground, sea, and air forces in the vital Mediterranean area will stand by his side until every objective in that bitter theater is attained.

Both of these new commanders will have American and British subordinate commanders whose names will be announced in a few days.

During the last two days at Teheran, Marshal Stalin, Mr. Churchill, and I looked ahead to the days and months and years that will follow Germany's defeat. We were united in determination that Germany must be stripped of her military might and be given no opportunity within the foreseeable future to regain that might.

The United Nations have no intention to enslave the German people. We wish them to have a normal chance to develop, in peace, as useful and respectable members of the European family. But we most certainly emphasize the word "respectable"—for we intend to rid them once and for all of Nazism and Prussian militarism and the fantastic and disastrous notion that they constitute the "master race."

We did discuss international relationships from the point of view of big, broad objectives, rather than details. But on the basis of what we did discuss, I can say even today that I do not think any insoluble differences will arise among Russia, Great Britain, and the United States.

In these conferences we were concerned with basic principles—principles which involve the security and the welfare and the standard of living of human beings in countries large and small.

To use an American and somewhat ungrammatical colloquialism, I may say that I "got along fine" with Marshal Stalin. He is a man who combines a tremendous, relentless determination with a stalwart good humor. I believe he is truly representative of the heart and soul of Russia; and I believe that we are going to get along very well with him and the Russian people—very well indeed.

Britain, Russia, China, and the United States and their allies represent more than three-quarters of the total population of the earth. As long as these four nations with great military power stick together in determination to keep the peace there will be no possibility of an aggressor nation arising to start another world war.

But those four powers must be united with and cooperate with all the freedom-loving peoples of Europe, and Asia, and Africa, and the Americas. The rights of every nation, large or small, must be respected and guarded as jealously as are the rights of every individual within our own Republic.

The doctrine that the strong shall dominate the weak is the doctrine of our enemies and we reject it.

But, at the same time, we are agreed that if force is necessary to keep international peace, international force will be applied for as long as it may be necessary.

It has been our steady policy—and it is certainly a common-sense policy—that the right of each nation to freedom must be measured by the willingness of that nation to fight for freedom. And today we salute our unseen allies in occupied countries—the underground resistance groups and the armies of liberation. They will provide potent forces against our enemies, when the day of the counterinvasion comes. . . .

The overwhelming majority of all the people in the world want peace. Most of them are fighting for the attainment of peace—not just a truce, not just an armistice—but peace that is as strongly enforced and as durable as mortal man can make it. If we are willing to fight for peace now, is it not good logic that we should use force if necessary, in the future, to keep the peace?

I believe, and I think I can say, that the other three great nations who are fighting so magnificently to gain peace are in complete agreement that we must be prepared to keep the peace by force. If the people of Germany and Japan are made to realize thoroughly that the world is not going to let them break out again, it is possible, and, I hope, probable, that they will abandon the philosophy of aggression—the belief that they can gain the whole world even at the risk of losing their own souls.

I shall have more to say about the Cairo and Teheran conferences when I make my report to the Congress in about two weeks' time. And, on that occasion, I shall also have a great deal to say about certain conditions here at home.

But today I wish to say that in all my travels, at home and abroad, it is the sight of our soldiers and sailors and their magnificent achievements which have given me the greatest inspiration and the greatest encouragement for the future.

To the members of our armed forces, to their wives, mothers, and fathers, I want to affirm the great faith and confidence that we have in General Marshall and in Admiral King, who direct all of our armed might throughout the world. Upon them falls the great responsibility of planning the strategy of determining where and when we shall fight. Both of these men have already gained high places in American history, which will record many evidences of their military genius that cannot be published today.

Some of our men overseas are now spending their third Christmas far from home. To them and to all others overseas or soon to go overseas, I can give assurance that it is the purpose of their government to win this war and to bring them home at the earliest possible time.

We here in the United States had better be sure that when our soldiers and sailors do come home they will find an America in which they are given full opportunities for education, and rehabilitation, social security, and employment and business enterprise under the free American system and that they will find a government which, by their votes as American citizens, they have had a full share in electing.

The American people have had every reason to know that this is a tough and destructive war. On my trip abroad, I talked with many military men who had faced our enemies in the field. These hardheaded realists testify to the strength and skill and resourcefulness of the enemy generals and men whom we must beat before final victory is won. The war is now reaching the stage where we shall all have to look forward to large casualty lists dead, wounded, and missing.

War entails just that. There is no easy road to victory. And the end is not yet in sight.

I have been back only for a week. It is fair that I

should tell you my impression. I think I see a tendency in some of our people here to assume a quick ending of the war—that we have already gained the victory. And, perhaps as a result of this false reasoning, I think I discern an effort to resume or even encourage an outbreak of partisan thinking and talking. I hope I am wrong. For, surely, our first and most foremost tasks are all concerned with winning the war and winning a just peace that will last for generations.

The massive offensives which are in the making—both in Europe and the Far East—will require every ounce of energy and fortitude that we and our allies can summon on the fighting fronts and in all the workshops at home. As I have said before, you cannot order up a great attack on a Monday and demand that it be delivered on Saturday.

Less than a month ago, I flew in a big army transport plane over the little town of Bethlehem, in Palestine.

Tonight, on Christmas Eve, all men and women everywhere who love Christmas are thinking of that ancient town and of the star of faith that shone there more than nineteen centuries ago.

American boys are fighting today in snow-covered mountains, in malarial jungles, on blazing deserts; they are fighting on the far stretches of the sea and above the clouds, and fighting for the thing for which they struggle. I think it is best symbolized by the message that came out of Bethlehem.

On behalf of the American people—your own people—I send this Christmas message to you who are in our armed forces:

In our hearts are prayers for you and for all your comrades in arms who fight to rid the world of evil.

We ask God's blessing upon you—upon your fathers, mothers, wives, and children, all your loved ones at home.

We ask that the comfort of God's grace shall be granted to those who are sick and wounded, and to those who are prisoners of war in the hands of the enemy, waiting for the day when they will again be free.

And we ask that God receive and cherish those who have given their lives, and that He keep them in honor and in the grateful memory of their countrymen forever.

God bless all of you who fight our battles on this Christmas Eve.

God bless us all. Keep us strong in our faith that we fight for a better day for humankind—here and everywhere.

In his eleventh message to Congress, January 11, 1944, Roosevelt continued to look forward as he laid out his Economic Bill of Rights.

THIS NATION IN the past two years has become an active partner in the world's greatest war against human slavery.

We have joined with like-minded people in order to defend ourselves in a world that has been gravely threatened with gangster rule.

But I do not think that any of us Americans can be content with mere survival. Sacrifices that we and our allies are making impose upon us all a sacred obligation to see to it that out of this war we and our children will gain something better than mere survival.

We are united in determination that this war shall not

be followed by another interim which leads to new disaster—that we shall not repeat the tragic errors of ostrich isolationism—that we shall not repeat the excesses of the wild twenties when this nation went for a joy ride on a roller coaster which ended in a tragic crash.

When Mr. Hull went to Moscow in October, and when I went to Cairo and Teheran in November, we knew that we were in agreement with our allies in our common determination to fight and win this war. But there were many vital questions concerning the future peace, and they were discussed in an atmosphere of complete candor and harmony.

In the last war such discussions, such meetings, did not even begin until the shooting had stopped and the delegates began to assemble at the peace table. There had been no previous opportunities for man-to-man discussions which lead to meetings of minds. The result was a peace which was not a peace.

That was a mistake which we are not repeating in this war.

And right here I want to address a word or two to some suspicious souls who are fearful that Mr. Hull or I have made "commitments" for the future which might pledge this nation to secret treaties, or to enacting the role of Santa Claus.

To such suspicious souls—using a polite terminology—I wish to say that Mr. Churchill, and Marshal Stalin, and Generalissimo Chiang Kai-shek are all thoroughly conversant with the provisions of our Constitution. And so is Mr. Hull. And so am I.

Of course we made some commitments. We most certainly committed ourselves to very large and very specific military plans which require the use of all Allied forces to bring about the defeat of our enemies at the earliest possible time.

But there were no secret treaties or political or financial commitments.

The one supreme objective for the future, which we discussed for each nation individually, and for all the United Nations, can be summed up in one word: security.

And that means not only physical security which provides safety from attacks by aggressors. It means also economic security, social security, moral security in a family of nations.

In the plain down-to-earth talks that I had with the Generalissimo and Marshal Stalin and Prime Minister Churchill, it was abundantly clear that they are all most deeply interested in the resumption of peaceful progress by their own peoples—progress toward a better life. All our allies want freedom to develop their lands and resources, to build up industry, to increase education and individual opportunity, and to raise standards of living.

All our allies have learned by bitter experience that real development will not be possible if they are to be diverted from their purpose by repeated wars—or even threats of war.

China and Russia are truly united with Britain and America in recognition of this essential fact:

The best interests of each nation, large and small, demand that all freedom-loving nations shall join together in a just and durable system of peace. In the present world situation, evidenced by the actions of Germany, Italy, and Japan, unquestioned military control over disturbers of the peace is as necessary among nations as it is among citizens in a community. And an equally basic essential to peace is a decent standard of living for all individual men and women and children in all nations. Freedom from fear is eternally linked with freedom from want. . . .

If ever there was a time to subordinate individual or group selfishness to the national good, that time is now. Disunity at home—bickerings, self-seeking partisanship, stoppages of work, inflation, business as usual, politics as usual, luxury as usual—these are the influences which can undermine the morale of the brave men ready to die at the front for us here.

Those who are doing most of the complaining are not deliberately striving to sabotage the national war effort. They are laboring under the delusion that the time is past when we must make prodigious sacrifices—that the war is already won and we can begin to slacken off. But the dangerous folly of that point of view can be measured by the distance that separates our troops from their ultimate objectives in Berlin and Tokyo and by the sum of all the perils that lie along the way.

Overconfidence and complacency are among our deadliest enemies. . . .

Therefore, in order to concentrate all our energies and resources on winning the war, and to maintain a fair and stable economy at home, I recommend that the Congress adopt:

1. A realistic tax law which will tax all unreasonable profits, both individual and corporate, and reduce the ultimate cost of the war to our sons and daughters. The tax bill now under consideration by the Congress does not begin to meet this test.

2. A continuation of the law for the renegotiation of war contracts—which will prevent exorbitant profits and assure fair prices to the government. For two long years

I have pleaded with the Congress to take undue profits out of war.

3. A cost-of-food law—which will enable the government (a) to place a reasonable floor under the prices the farmer may expect for his production, and (b) to place a ceiling on the prices a consumer will have to pay for the food he buys. This should apply to necessities only; and will require public funds to carry out. It will cost in appropriations about one percent of the present annual cost of the war.

4. Early reenactment of the stabilization statute of October 1942. This expires June 30, 1944, and if it is not extended well in advance, the country might just as well expect price chaos by summer. We cannot have stabilization by wishful thinking. We must take positive action to maintain the integrity of the American dollar.

5. A national-service law—which, for the duration of the war, will prevent strikes, and, with certain appropriate exceptions, will make available for war production or for any other essential services every able-bodied adult in this nation.

These five measures together form a just and equitable whole. I would not recommend a national-service law unless the other laws were passed to keep down the cost of living, to share equitably the burdens of taxation, to hold the stabilization line, and to prevent undue profits. . . .

As you know, I have for three years hesitated to recommend a national-service act. Today, however, I am convinced of its necessity. Although I believe that we and our allies can win the war without such a measure, I am certain that nothing less than total mobilization of all our resources of manpower and capital will guarantee an earlier victory, and reduce the toll of suffering and sorrow and blood. . . .

It is our duty now to begin to lay the plans and determine the strategy for the winning of a lasting peace and the establishment of an American standard of living higher than ever before known. We cannot be content, no matter how high that general standard of living may be, if some fraction of our people—whether it be one-third or one-fifth, or one-tenth—is ill-fed, ill-clothed, ill-housed, and insecure.

This Republic had its beginning, and grew to its present strength, under the protection of certain inalienable political rights—among them the right of free speech, free press, free worship, trial by jury, freedom from unreasonable searches and seizures. They were our rights to life and liberty.

As our nation has grown in size and stature, however—as our industrial economy expanded—these political rights proved inadequate to assure us equality in the pursuit of happiness.

We have come to a clear realization of the fact that true individual freedom cannot exist without economic security and independence. "Necessitous men are not free men." People who are hungry and out of a job are the stuff of which dictatorships are made.

In our day these economic truths have become accepted as self-evident. We have accepted, so to speak, a second Bill of Rights under which a new basis of security and prosperity can be established for all—regardless of station, race, or creed.

Among these are:

The right to a useful and remunerative job in the industries or shops or farms or mines of the nation;

The right to earn enough to provide adequate food and clothing and recreation;

The right of every farmer to raise and sell his products at a return which will give him and his family a decent living;

The right of every businessman, large and small, to trade in an atmosphere of freedom from unfair competition and domination by monopolies at home or abroad;

The right of every family to a decent home;

The right to adequate medical care and the opportunity to achieve and enjoy good health;

The right to adequate protection from the economic fears of old age, sickness, accident, and unemployment;

The right to a good education.

All of these rights spell security. And after this war is won we must be prepared to move forward, in the implementation of these rights, to new goals of human happiness and well-being. . . .

I ask the Congress to explore the means for implementing this economic Bill of Rights—for it is definitely the responsibility of the Congress so to do. Many of these problems are already before committees of the Congress in the form of proposed legislation. I shall from time to time communicate with the Congress with respect to these and further proposals. In the event that no adequate program of progress is evolved, I am certain that the nation will be conscious of the fact.

Our fighting men abroad—and their families at home—expect such a program and have the right to insist upon it. It is to their demands that this government should pay heed rather than to the whining demands of selfish pressure groups who seek to feather their nests while young Americans are dying.

The foreign policy that we have been following—the policy that guided us at Moscow, Cairo, and Teheran—is based on the common sense principle which was best expressed by Benjamin Franklin on July 4, 1776: "We must all hang together, or assuredly we shall all hang separately."

I have often said that there are no two fronts for America in this war. There is only one front. There is one line of unity which extends from the hearts of the people at home to the men of our attacking forces in our farthest outposts. When we speak of our total effort, we speak of the factory and the field and the mine as well as of the battleground—we speak of the soldier and the civilian, the citizen and his government.

Each and every one of us has a solemn obligation under God to serve this nation in its most critical hour—to keep this nation great—to make this nation greater in a better world.

Historians have yet to agree on what Roosevelt knew about the Holocaust, when he knew what he knew—and whether he could have done more to stop it. His first public acknowledgement came on March 24, 1944, when he issued this strong statement condemning German and Japanese war crimes.

THE UNITED NATIONS are fighting to make a world in which tyranny and aggression cannot exist; a world based on freedom, equality, and justice; a world in which all persons regardless of race, color, or creed may live in peace, honor, and dignity.

In the meantime in most of Europe and parts of Asia,

the systematic torture and murder of civilians—men, women, and children—by the Nazis and the Japanese continue unabated. In areas subjugated by the aggressors, innocent Poles, Czechs, Norwegians, Dutch, Danes, French, Greeks, Russians, Chinese, Filipinos—and many others—are being starved or frozen to death or murdered in cold blood in a campaign of savagery.

The slaughters of Warsaw, Lidice, Kharkov, and Nanking—the brutal torture and murder by the Japanese, not only of civilians, but of our own gallant American soldiers and fliers—these are startling examples of what goes on day by day, year in and year out, wherever the Nazis and the Japs are in military control—free to follow their barbaric purpose.

In one of the blackest crimes of all history begun by the Nazis in the day of peace and multiplied by them a hundred times in time of war the wholesale systematic murder of the Jews of Europe goes on unabated every hour. As a result of the events of the last few days hundreds of thousands of Jews, who while living under persecution have at least found a haven from death in Hungary and the Balkans, are now threatened with annihilation as Hitler's forces descend more heavily upon these lands. That these innocent people, who have already survived a decade of Hitler's fury, should perish on the very eve of triumph over the barbarism which their persecution symbolizes would be a major tragedy.

It is therefore fitting that we should again proclaim our determination that none who participate in these acts of savagery shall go unpunished. The United Nations have made it clear that they will pursue the guilty and deliver them up in order that justice be done. That warning applies not only to the leaders but also to their functionaries and subordinates in Germany and in the satellite countries. All who knowingly take part in the deportation of Jews to their death in Poland or Norwegians and French to their death in Germany are equally guilty with the executioner. All who share the guilt shall share the punishment.

Hitler is committing these crimes against humanity in the name of the German people. I ask every German and every man everywhere under Nazi domination to show the world by his action that in his heart he does not share these insane criminal desires. Let him hide these pursued victims, help them to get over their borders, and do what he can to save them from the Nazi hangman. I ask him also to keep watch, and to record the evidence that will one day be used to convict the guilty.

In the meantime, and until the victory that is now assured is won, the United States will persevere in its efforts to rescue the victims of brutality of the Nazis and the Japs. Insofar as the necessity of military operations permit, this government will use all means at its command to aid the escape of all intended victims of the Nazi and Jap executioner regardless of race or religion or color. We call upon the free peoples of Europe and Asia temporarily to open their frontiers to all victims of oppression. We shall find havens of refuge for them, and we shall find the means for their maintenance and support until the tyrant is driven from their homelands and they may return.

In the name of justice and humanity let all freedom-loving people rally to this righteous undertaking.

Before there was Checkers, there was Fala. In the 1944 campaign, FDR's Scottie, his inseparable companion, provided the basis for some very welcome drollery: "These Republican leaders have not been content with attacks—on me, or my wife, or on my sons. No, not content with that, they now include my little dog, Fala. Well, of course, I don't resent attacks, and my family doesn't resent attacks, but Fala does *resent them. You know—you know—Fala's Scotch, and being a Scottie, as soon as he learned that the Republican fiction writers in Congress and out had concocted a story that I had left him behind on an Aleutian Island and had sent a destroyer back to find him—at a cost to the taxpayers of two or three, or eight or twenty million dollars—his Scotch soul was furious. He has not been the same dog since."*

Roosevelt's Fireside Chat of June 5, 1944, gave the American people some crucial—and, for a change—good news: The Allies had captured Rome.

YESTERDAY, ON JUNE 4, 1944, Rome fell to American and Allied troops. The first of the Axis capitals is now in our hands. One up and two to go!

It is perhaps significant that the first of these capitals to fall should have the longest history of all of them. The story of Rome goes back to the time of the foundations of our civilization. We can still see there monuments of the time when Rome and the Romans controlled the whole of the then known world. That, too, is significant, for the United Nations are determined that in the future no one city and no one race will be able to control the whole of the world.

In addition to the monuments of the older times, we also see in Rome the great symbol of Christianity, which has reached into almost every part of the world. There are other shrines and other churches in many places, but the churches and shrines of Rome are visible symbols of the faith and determination of the early saints and martyrs that Christianity should live and become universal. And tonight it will be a source of deep satisfaction that the freedom of the pope and the Vatican City is assured by the armies of the United Nations.

It is also significant that Rome has been liberated by the armed forces of many nations. The American and British armies—who bore the chief burdens of battle—found at their sides our own North American neighbors, the gallant Canadians. The fighting New Zealanders from the far South Pacific, the courageous French and the French Moroccans, the South Africans, the Poles, and the East Indians—all of them fought with us on the bloody approaches to the city of Rome.

The Italians, too, forswearing a partnership in the Axis which they never desired, have sent their troops to join us in our battles against the German trespassers on their soil.

The prospect of the liberation of Rome meant enough to Hitler and his generals to induce them to fight desperately at great cost of men and materials and with great sacrifice to their crumbling eastern line and to their western front. No thanks are due to them if Rome was spared the devastation which the Germans wreaked on Naples and other Italian cities. The Allied generals maneuvered so skillfully that the Nazis could only have stayed long enough to damage Rome at the risk of losing their armies.

But Rome is of course more than a military objective.

Ever since before the days of the Caesars, Rome has stood as a symbol of authority. Rome was the Republic. Rome was the Empire. Rome was and is in a sense the Catholic Church, and Rome was the capital of a United Italy. Later, unfortunately, a quarter of a century ago,

Rome became the seat of Fascism—one of the three capitals of the Axis.

For this quarter century the Italian people were enslaved. They were degraded by the rule of Mussolini from Rome. They will mark its liberation with deep emotion. In the north of Italy, the people are still dominated and threatened by the Nazi overlords and their Fascist puppets. . . .

Our victory comes at an excellent time, while our Allied forces are poised for another strike at Western Europe—and while the armies of other Nazi soldiers nervously await our assault. And in the meantime our gallant Russian allies continue to make their power felt more and more.

From a strictly military standpoint, we had long ago accomplished certain of the main objectives of our Italian campaign—the control of the major islands, the control of the sea lanes of the Mediterranean to shorten our combat and supply lines, and the capture of the airports of Foggia, south of Rome, from which we have struck telling blows on the continent—the whole of the continent all the way up to the Russian front.

It would be unwise to inflate in our own minds the military importance of the capture of Rome. We shall have to push through a long period of greater effort and fiercer fighting before we get into Germany itself. The Germans have retreated thousands of miles, all the way from the gates of Cairo, through Libya and Tunisia and Sicily and southern Italy. They have suffered heavy losses, but not great enough yet to cause collapse.

Germany has not yet been driven to surrender. Germany has not yet been driven to the point where she will be unable to recommence world conquest a generation hence.

Therefore, the victory still lies some distance ahead. That distance will be covered in due time—have no fear of that. But it will be tough and it will be costly, as I have told you many, many times.

In Italy the people had lived so long under the corrupt rule of Mussolini that, in spite of the tinsel at the top . . . their economic condition had grown steadily worse. Our troops have found starvation, malnutrition, disease, a deteriorating education, and lowered public health—all by-products of the Fascist misrule.

The task of the Allies in occupation has been stupendous. We have had to start at the very bottom, assisting local governments to reform on democratic lines. We have had to give them bread to replace that which was stolen out of their mouths by the Germans. We have had to make it possible for the Italians to raise and use their own local crops. We have to help them cleanse their schools of Fascist trappings.

I think the American people as a whole approve the salvage of these human beings, who are only now learning to walk in a new atmosphere of freedom.

Some of us may let our thoughts run to the financial cost of it. Essentially it is what we can call a form of relief. And at the same time, we hope that this relief will be an investment for the future—an investment that will pay dividends by eliminating Fascism, by ending any Italian desires to start another war of aggression in the future. And that means that they are dividends which justify such an investment, because they are additional supports for world peace.

The Italian people are capable of self-government. We do not lose sight of their virtues as a peace-loving nation.

We remember the many centuries in which the Italians were leaders in the arts and sciences, enriching the lives of all mankind.

We remember the great sons of the Italian people—Galileo and Marconi, Michelangelo and Dante—and that fearless discoverer who typifies the courage of Italy, Christopher Columbus.

Italy cannot grow in stature by seeking to build up a great militaristic empire. Italians have been overcrowded within their own territories, but they do not need to try to conquer the lands of other peoples in order to find the breath of life. Other peoples may not want to be conquered.

In the past, Italians have come by the millions into the United States. They have been welcomed, they have prospered, they have become good citizens, community and governmental leaders. They are not Italian-Americans. They are Americans—Americans of Italian descent.

The Italians have gone in great numbers to the other Americas—Brazil and the Argentine, for example—hundreds and hundreds of thousands of them. They have gone to many other nations in every continent of the world, giving of their industry and their talents, and achieving success and the comfort of good living, and good citizenship.

Italy should go on as a great mother nation, contributing to the culture and the progress and the goodwill of all mankind—developing her special talents in the arts and crafts and sciences, and preserving her historic and cultural heritage for the benefit of all peoples.

We want and expect the help of the future Italy

toward lasting peace. All the other nations opposed to Fascism and Nazism ought to help to give Italy a chance.

The Germans, after years of domination in Rome, left the people in the Eternal City on the verge of starvation. We and the British will do and are doing everything we can to bring them relief. Anticipating the fall of Rome, we made preparations to ship food supplies to the city, but, of course, it should be borne in mind that the needs are so great, the transportation requirements of our armies so heavy, that improvement must be gradual. But we have already begun to save the lives of the men, women, and children of Rome.

This, I think, is an example of the efficiency of our machinery of war. The magnificent ability and energy of the American people in growing the crops, building the merchant ships, in making and collecting the cargoes, in getting the supplies over thousands of miles of water, and thinking ahead to meet emergencies—all this spells, I think, an amazing efficiency on the part of our armed forces, all the various agencies working with them, and American industry and labor as a whole.

No great effort like this can be a hundred percent perfect, but the batting average is very, very high.

And so I extend the congratulations and thanks tonight of the American people to General Alexander, who has been in command of the whole Italian operation; to our General Clark and General Leese of the Fifth and the Eighth armies; to General Wilson, the supreme allied commander of the Mediterranean theater; to General Devers, his American deputy; to General Eaker; to Admirals Cunningham and Hewitt; and to all their brave officers and men.

May God bless them and watch over them and over all of our gallant, fighting men.

During his final inaugural ceremony on January 20, 1945, Roosevelt was particularly lofty, an affect that, in retrospect, hardly seems inappropriate.

❧ MR. CHIEF JUSTICE, Mr. Vice President, My Friends: You will understand and, I believe, agree with my wish that the form of this inauguration be simple and its words brief.

We Americans of today, together with our allies, are passing through a period of supreme test. It is a test of our courage—of our resolve—of our wisdom—of our essential democracy.

If we meet that test—successfully and honorably—we shall perform a service of historic importance which men and women and children will honor throughout all time.

As I stand here today, having taken the solemn oath of office in the presence of my fellow countrymen—in the presence of our God—I know that it is America's purpose that we shall not fail.

In the days and the years that are to come, we shall work for a just and honorable peace, a durable peace, as today we work and fight for total victory in war.

We can and we will achieve such a peace.

We shall strive for perfection. We shall not achieve it immediately, but we still shall strive. We may make mistakes, but they must never be mistakes which result from faintness of heart or abandonment of moral principle.

I remember that my old schoolmaster, Dr. Peabody,

said—in days that seemed to us then to be secure and untroubled: "Things in life will not always run smoothly. Sometimes we will be rising toward the heights—then all will seem to reverse itself and start downward. The great fact to remember is that the trend of civilization itself is forever upward; that a line drawn through the middle of the peaks and the valleys of the centuries always has an upward trend."

Our Constitution of 1787 was not a perfect instrument; it is not perfect yet. But it provided a firm base upon which all manner of men, of all races and colors and creeds, could build our solid structure of democracy.

Today, in this year of war, 1945, we have learned lessons—at a fearful cost—and we shall profit by them.

We have learned that we cannot live alone, at peace; that our own well-being is dependent on the well-being of other nations, far away. We have learned that we must live as men and not as ostriches, nor as dogs in the manger.

We have learned to be citizens of the world, members of the human community.

We have learned the simple truth, as Emerson said, that "the only way to have a friend is to be one."

We can gain no lasting peace if we approach it with suspicion and mistrust—or with fear. We can gain it only if we proceed with the understanding and the confidence and the courage which flow from conviction.

The Almighty God has blessed our land in many ways. He has given our people stout hearts and strong arms with which to strike mighty blows for freedom and truth. He has given to our country a faith which has become the hope of all peoples in an anguished world.

So we pray to Him now for the vision to see our way

clearly—to see the way that leads to a better life for ourselves and for all our fellow men—to the achievement of His will to peace on earth.

In his March 1 report to a joint session of Congress on the Yalta Conference, the last speech he ever made, Roosevelt addressed those gathered from a wheelchair, and sounded completely exhausted, although his words were as strong as ever.

I HOPE THAT you will pardon me for this unusual posture of sitting down during the presentation of what I want to say, but I know that you will realize that it makes it a lot easier for me not to have to carry about ten pounds of steel around on the bottom of my legs; and also because of the fact that I have just completed a fourteen-thousand-mile trip.

First of all, I want to say, it is good to be home.

It has been a long journey. I hope you will also agree that it has been, so far, a fruitful one.

Speaking in all frankness, the question of whether it is entirely fruitful or not lies to a great extent in your hands. For unless you here in the halls of the American Congress—with the support of the American people—concur in the general conclusions reached at Yalta, and give them your active support, the meeting will not have produced lasting results.

That is why I have come before you at the earliest hour I could after my return. I want to make a personal report to you—and, at the same time, to the people of the country. Many months of earnest work are ahead of us all, and I should like to feel that when the last stone is laid on the structure of international peace, it will be an achievement for which all of us in America have worked steadfastly and unselfishly—together.

I am returning from this trip—that took me so far—refreshed and inspired. I was well the entire time. I was not ill for a second, until I arrived back in Washington, and there I heard all of the rumors which had occurred in my absence. I returned from the trip refreshed and inspired. The Roosevelts are not, as you may suspect, averse to travel. We seem to thrive on it!

Far away as I was, I was kept constantly informed of affairs in the United States. The modern miracles of rapid communication have made this world very small. We must always bear in mind that fact, when we speak or think of international relations. I received a steady stream of messages from Washington—I might say from not only the executive branch with all its departments, but also from the legislative branch—and except where radio silence was necessary for security purposes, I could continuously send messages any place in the world. And of course, in a grave emergency, we could have even risked the breaking of the security rule.

I come from the Crimea Conference with a firm belief that we have made a good start on the road to a world of peace.

There were two main purposes in this Crimea Conference. The first was to bring defeat to Germany with the greatest possible speed, and the smallest possible loss of Allied men. That purpose is now being carried out in great force. The German army, and the German people, are feeling the ever-increasing might of our fighting men and of the Allied armies. Every hour gives us added pride in the heroic advance of our troops in Germany on German soil toward a meeting with the gallant Red Army.

The second purpose was to continue to build the foundation for an international accord that would bring order and security after the chaos of the war, that would give some assurance of lasting peace among the nations of the world.

Toward that goal also, a tremendous stride was made.

At Teheran, a little over a year ago, there were long-range military plans laid by the Chiefs of Staff of the three most powerful nations. Among the civilian leaders at Teheran, however, at that time, there were only exchanges of views and expressions of opinion. No political arrangements were made and none was attempted.

At the Crimea Conference, however, the time had come for getting down to specific cases in the political field.

There was on all sides at this conference an enthusiastic effort to reach an agreement. Since the time of Teheran, a year ago, there had developed among all of us—what shall I call it?—a greater facility in negotiating with each other, that augurs well for the peace of the world. We know each other better.

I have never for an instant wavered in my belief that an agreement to insure world peace and security can be reached.

When we met at Yalta, in addition to laying our strategic and tactical plans for the complete and final military victory over Germany, there were other problems of vital political consequence.

For instance, first, there were the problems of the occupation and control of Germany after victory—the complete destruction of her military power, and the assurance that neither the Nazis nor Prussian militarism could again be revived to threaten the peace and the civilization of the world.

Second—again for example—there was the settlement of the few differences that remained among us with respect to the International Security Organization after the Dumbarton Oaks Conference. As you remember, at that time, I said that we had agreed 90 percent. Well, that's a pretty good percentage. I think the other 10 percent was ironed out at Yalta.

Third, there were the general political and economic problems common to all of the areas which had been or would be liberated from the Nazi yoke. This is a very special problem. We over here find it difficult to understand the ramifications of many of these problems in foreign lands, but we are trying to.

Fourth, there were the special problems created by a few instances such as Poland and Yugoslavia.

Days were spent in discussing these momentous matters and we argued freely and frankly across the table. But at the end, on every point, unanimous agreement was reached. And more important even than the agreement of words, I may say we achieved a unity of thought and a way of getting along together.

Of course, we know that it was Hitler's hope—and the German warlords'—that we would not agree, that some slight crack might appear in the solid wall of Allied unity, a crack that would give him and his fellow gangsters one last hope of escaping their just doom. That is the objective for which his propaganda machine has been working for many months.

But Hitler has failed.

Never before have the major Allies been more closely

united not only in their war aims but also in their peace aims. And they are determined to continue to be united with each other and with all peace-loving nations so that the ideal of lasting peace will become a reality.

The Soviet, British, and United States Chiefs of Staff held daily meetings with each other. They conferred frequently with Marshal Stalin and with Prime Minister Churchill and with me, on the problems of coordinating the strategic and tactical efforts of the Allied powers. They completed their plans for the final knockout blows to Germany.

At the time of the Teheran Conference, the Russian front was removed so far from the American and British fronts that, while certain long-range strategic cooperation was possible, there could be no tactical, day-by-day coordination. They were too far apart. But Russian troops have now crossed Poland. They are fighting on the eastern soil of Germany herself; British and American troops are now on German soil close to the Rhine River in the west. It is a different situation today from what it was 14 months ago; a closer tactical liaison has become possible for the first time in Europe—and, in the Crimea Conference, that was something else that was accomplished.

Provision was made for daily exchange of information between the armies under the command of General Eisenhower on the western front, and those armies under the command of the Soviet marshals on that long eastern front, and also with our armies in Italy—without the necessity of going through the Chiefs of Staff in Washington or London as in the past.

You have seen one result of this exchange of information in the recent bombings by American and English aircraft of points which are directly related to the Russian advance on Berlin.

We made it clear again at Yalta, and I now repeat, that unconditional surrender does not mean the destruction or enslavement of the German people. The Nazi leaders have deliberately withheld that part of the Yalta declaration from the German press and radio. They seek to convince the people of Germany that the Yalta declaration does mean slavery and destruction for them; they are working at it day and night, for that is how the Nazis hope to save their own skins, and deceive their people into continued and useless resistance.

We did, however, make it clear at the conference just what unconditional surrender does mean for Germany.

It means the temporary control of Germany by Great Britain, Russia, France, and the United States. Each of these nations will occupy and control a separate zone of Germany—and the administration of the four zones will be coordinated in Berlin by a Control Council composed of representatives of the four nations.

Unconditional surrender means something else. It means the end of Nazism. It means the end of the Nazi party and of all its barbaric laws and institutions.

It means the termination of all militaristic influence in the public, private, and cultural life of Germany.

It means for the Nazi war criminals a punishment that is speedy and just—and severe.

It means the complete disarmament of Germany; the destruction of its militarism and its military equipment; the end of its production of armament; the dispersal of all its armed forces; the permanent dismemberment of the German General Staff which has so often shattered the peace of the world.

It means that Germany will have to make reparations in kind for the damage which has been done to the innocent victims of its aggression.

As the Allied armies have marched to military victory, they have liberated people whose liberties had been crushed by the Nazis for four long years, whose economy has been reduced to ruin by Nazi despoilers.

There have been instances of political confusion and unrest in these liberated areas that is not unexpected, as in Greece or in Poland or in Yugoslavia, and there may be more. Worse than that, there actually began to grow up in some of these places queer ideas of, for instance, "spheres of influence" that were incompatible with the basic principles of international collaboration. If allowed to go on unchecked, these developments might have had tragic results in time.

We met in the Crimea, determined to settle this matter of liberated areas. Things that might happen that we cannot foresee at this moment might happen suddenly—unexpectedly—next week or next month. And I am happy to confirm to the Congress that we did arrive at a settlement and, incidentally, a unanimous settlement.

The three most powerful nations have agreed that the political and economic problems of any area liberated from Nazi conquest, or of any former Axis satellite, are a joint responsibility of all three governments. They will join together, during the temporary period of instability after hostilities to help the people of any liberated area, or of any former satellite state, to solve their own problems through firmly established democratic processes.

They will endeavor to see to it that the people who carry on the interim government between occupation of Germany and true independence will be as representative as possible of all democratic elements in the population, and that free elections are held as soon as possible thereafter.

Responsibility for political conditions thousands of miles away can no longer be avoided by this great nation. Certainly, I do not want to live to see another war. As I have said, the world is smaller every year. The United States now exerts a tremendous influence in the cause of peace throughout all the world. What we people over here are thinking and talking about is in the interest of peace, because it is known all over the world. The slightest remark in either house of the Congress is known all over the world the following day. We will continue to exert that influence, only if we are willing to continue to share in the responsibility for keeping the peace. It will be our own tragic loss, I think, if we were to shirk that responsibility.

The final decisions in these areas are going to be made jointly; and therefore they will often be a result of give-and-take compromise. The United States will not always have its way 100 percent nor will Russia nor Great Britain. We shall not always have ideal answers—solutions to complicated international problems, even though we are determined continuously to strive toward that ideal. But I am sure that under the agreements reached at Yalta, there will be a more stable political Europe than ever before.

One outstanding example of joint action by the three major Allied powers in the liberated areas was the solution reached on Poland. The whole Polish question was a potential source of trouble in postwar Europe—as it has been sometimes before—and we came to the conference determined to find a common ground for its solution. And we did—even though everybody does not agree with us, obviously.

Our objective was to help to create a strong, independent, and prosperous nation. That is the thing we must always remember, those words, agreed to by Russia, by Britain, and by the United States: the objective of making Poland a strong, independent, and prosperous nation, with a government ultimately to be selected by the Polish people themselves.

To achieve that objective, it was necessary to provide for the formation of a new government much more representative than had been possible while Poland was enslaved. There were, as you know, only two governments: one in London, one in Lublin—practically in Russia. Accordingly, steps were taken at Yalta to reorganize the existing provisional government in Poland on a broader democratic basis, so as to include democratic leaders now in Poland and those abroad. This new, reorganized government will be recognized by all of us as the temporary government of Poland. Poland needs a temporary government in the worst way—an ad interim government, I think is another way of putting it.

However, the new Polish Provisional Government of National Unity will be pledged to holding a free election as soon as possible on the basis of universal suffrage and a secret ballot.

Throughout history, Poland has been the corridor through which attacks on Russia have been made. Twice in this generation, Germany has struck at Russia through this corridor. To insure European security and world peace, a strong and independent Poland is necessary to prevent that from happening again.

The decision with respect to the boundaries of Poland

was, frankly, a compromise. I did not agree with all of it, by any means, but we did not go as far as Britain wanted, in certain areas; we did not go as far as Russia wanted, in certain areas; and we did not go so far as I wanted, in certain areas. It *was* a compromise. The decision is one, however, under which the Poles will receive compensation in territory in the north and west in exchange for what they lose by the Curzon Line in the east. The limits of the western border will be permanently fixed in the final peace conference. We know, roughly, that it will include—in the new, strong Poland—quite a large slice of what now is called Germany. And it was agreed, also, that the new Poland will have a large and long coastline, and many new harbors. Also, that most of East Prussia will go to Poland. A corner of it will go to Russia. Also, that the anomaly of the Free State of Danzig will come to an end; I think Danzig would be a lot better if it were Polish.

It is well known that the people east of the Curzon Line—just for example, here is why I compromised—are predominantly White Russian and Ukrainian—they are not Polish; and a very great majority of the people west of the line are predominantly Polish, except in that part of East Prussia and eastern Germany, which will go to the new Poland.

I am convinced that the agreement on Poland, under the circumstances, is the most hopeful agreement possible for a free, independent, and prosperous Polish state.

The Crimea Conference was a meeting of the three major military powers on whose shoulders rested chief responsibility—and burden of the war. Although, for this reason, France was not a participant in the conference, no one should detract from the recognition that was accorded there of her role in the future of Europe and the future of the world.

France has been invited to accept a zone of control in Germany, and to participate as a fourth member of the Allied Control Council of Germany.

She has been invited to join as a sponsor of the International Conference at San Francisco next month.

She will be a permanent member of the International Security Council together with the other four major powers.

And, finally, we have asked that France be associated with us in our joint responsibility over all the liberated areas of Europe.

Agreement was reached on Yugoslavia, as announced in the communiqué; and we hope that it is in process of fulfillment. But, not only there but in some other places, we have to remember that there are a great many prima donnas in the world. All of them wish to be heard before anything becomes final, so we may have a little delay while we listen to more prima donnas.

Quite naturally, this conference concerned itself only with the European war and with the political problems of Europe—and not with the Pacific war.

In Malta, however, our combined British and American staffs made their plans to increase the attack against Japan.

The Japanese warlords know that they are not being overlooked. They have felt the force of our B-29s, and our carrier planes; they have felt the naval might of the United States, and do not appear very anxious to come out and try it again. . . .

It is still a long, tough road to Tokyo. It is longer to go to Tokyo than it is to Berlin, in every sense of the word. The defeat of Germany will not mean the end of the war against Japan. On the contrary, we must be prepared for a long and costly struggle in the Pacific.

But the unconditional surrender of Japan is as essential as the defeat of Germany. I say that advisedly, with the thought in mind that that is especially true if our plans for world peace are to succeed. For Japanese militarism must be wiped out as thoroughly as German militarism.

On the way back from the Crimea, I made arrangements to meet personally King Farouk of Egypt; Haile Selassie, the emperor of Ethiopia; and King Ibn Saud of Saudi Arabia. Our conversations had to do with matters of common interest. They will be of great mutual advantage because they gave me, and a good many of us, an opportunity of meeting and talking face to face, and of exchanging views in personal conversation instead of formal correspondence.

For instance, on the problem of Arabia, I learned more about that whole problem—the Moslem problem, the Jewish problem—by talking with Ibn Saud for five minutes, than I could have learned in the exchange of two or three dozen letters.

On my voyage, I had the benefit of seeing the army and navy and the air force at work.

All Americans, I think, would feel as proud of our armed forces as I am, if they could see and hear what I saw and heard.

Against the most efficient professional soldiers and sailors and airmen of all history, our men stood and fought and won.

This is our chance to see to it that the sons and the grandsons of these gallant fighting men do not have to do it all over again in a few years. . ..

The Crimea Conference was a successful effort by the three leading nations to find a common ground for peace. It ought to spell the end of the system of unilateral action, the exclusive alliances, the spheres of influence, the balances of power, and all the other expedients that have been tried for centuries and have always failed.

We propose to substitute for all these a universal organization in which all peace-loving nations will finally have a chance to join.

I am confident that the Congress and the American people will accept the results of this conference as the beginnings of a permanent structure of peace upon which we can begin to build, under God, that better world in which our children and grandchildren—yours and mine, the children and grandchildren of the whole world—must live, and can live.

And that, my friends, is the principal message I can give you. But I feel it very deeply, as I know that all of you are feeling it today, and are going to feel it in the future.

The last words he wrote, the night before he died, in his Jefferson Day speech, captured perfectly the spirit of all he had said and written in his illustrious career. "To you, and to all Americans who dedicate themselves with us to the making of an abiding peace, I say: The only limit to our realization of tomorrow will be our doubts of today. Let us move forward with strong and active faith."

Sources

Alsop, Joseph. *FDR: A Centenary Remembrance* (Viking, 1982).

The Essential Franklin Delano Roosevelt, "Library of Freedom" (Gramercy Books, 1995).

Freedman, Russell. *Franklin Delano Roosevelt* (Clarion Books, 1990).

Lash, Joseph P. *Eleanor and Franklin* (Norton, 1971).

———. *Roosevelt and Churchill* (Norton, 1976).

The Public Papers and Addresses of Franklin D. Roosevelt, 1937–1940 (Macmillan, 1941).